WHEN ROBOTS KILL

WHEN ROBOTS KILL

ARTIFICIAL INTELLIGENCE UNDER CRIMINAL LAW

GABRIEL HALLEVY

NORTHEASTERN UNIVERSITY PRESS · Boston

Northeastern University Press
An imprint of University Press of New England
www.upne.com
© 2013 Northeastern University
All rights reserved
Manufactured in the United States of America
Designed by Vicki Kuskowski
Typeset in Melior by Copperline Book Services, Inc.

University Press of New England is a member of the Green Press Initiative. The paper used in this book meets their minimum requirement for recycled paper.

For permission to reproduce any of the material in this book, contact Permissions, University Press of New England, One Court Street, Suite 250, Lebanon NH 03766; or visit www.upne.com

Library of Congress Cataloging-in-Publication Data
Halevi, Gavri'el.
When robots kill : artificial intelligence under criminal law / Gabriel Hallevy.
pages cm.
Includes bibliographical references and index.
ISBN 978–1–55553–801–9 (cloth) —
ISBN 978–1–55553–805–7 (pbk.) —
ISBN 978–1–55553–806–4 (ebook) (print)
1. Criminal liability. 2. Artificial intelligence—Law and legislation—Criminal provisions. I. Title.
K5070.H35 2013
345'.0252—dc23   2012033518

5 4 3 2 1

TO MY DEAR FAMILY

And God said, let us make man in our image, after our
likeness: and let them have dominion over the fish of the sea,
and over the fowl of the air, and over the cattle,
and over all the earth, and over every creeping thing that
creepeth upon the earth. So God created man in
his own image, in the image of God created he him;
male and female created he them.
—GENESIS 1:26–27

CONTENTS

List of Tables · xiii
Preface · xv

1

THE EMERGENCE OF *MACHINA SAPIENS CRIMINALIS* · 1

1.1. The Endless Quest for *Machina Sapiens* · 1

1.1.1. History and Prehistory of Artificial Intelligence · 1

1.1.2. Defining Artificial Intelligence, and the Endlessness of the Quest for *Machina Sapiens* · 4

1.1.3. Turning Disadvantages into Advantages: Industrial and Private Use of Artificial Intelligence · 12

1.2. On Evolution and Devolution: *Machina Sapiens Criminalis* · 14

1.2.1. Serve and Protect: Human Fears of Human-Robot Coexistence · 14

1.2.2. Evolution and Devolution: *Machina Sapiens Criminalis* as a By-product · 17

1.2.3. Inapplicability of the Zoological Legal Model to AI Technology · 22

1.3. The Modern Offender · 25

1.3.1. Basic Requirements for a Given Offense · 26

1.3.2. Requirements of a Given Offense: Modern Criminal Liability as a Matrix of Minimalism · 29

1.3.3. The Case of Nonhuman Corporations (Round 1) · 34

2

AI CRIMINAL LIABILITY FOR INTENTIONAL OFFENSES · 38

2.1. The Factual Element Requirement · 38

2.1.1. Conduct · 39

2.1.2. Circumstances · 42
2.1.3. Results and Causal Connection · 43
2.2. The Mental Element Requirement · 44
2.2.1. Structure of the *Mens Rea* Requirement · 45
2.2.2. Fulfillment of the Cognitive Aspect · 49
2.2.3. Fulfillment of the Volitive Aspect · 56
2.3. Criminally Liable Entities for Intentional AI Offenses · 64
2.3.1. AI Entity Liability · 64
2.3.2. Human Liability: Perpetration-through-Another · 69
2.3.3. Joint Human and AI Entity Liability: Probable Consequence Liability · 75
Closing the Opening Example: Intentional Killing Robot · 82

3

AI CRIMINAL LIABILITY FOR NEGLIGENCE OFFENSES · 84

3.1. The Factual Element Requirement · 85
3.2. The Mental Element Requirement · 85
3.2.1. Structure of Negligence Requirement · 85
3.2.2. Fulfillment of the Negligence Requirement: Is Objectivity Subjective? · 90
3.3. Criminally Liable Entities for AI Negligence Offenses · 95
3.3.1. AI Entity Liability · 96
3.3.2. Human Liability: Perpetration-through-Another and Semi-Innocent Agents · 97
3.3.3. Joint Human and AI Entity Liability: Probable Consequence Liability · 100
Closing the Opening Example: Negligent Killing Robot · 101

4

AI CRIMINAL LIABILITY FOR STRICT LIABILITY OFFENSES · 104

4.1. The Factual Element Requirement · 105
4.2. The Mental Element Requirement · 105
4.2.1. Structure of Strict Liability Requirement · 105
4.2.2. Fulfillment of Strict Liability: Making the Factual Become Mental · 109

4.3. Criminally Liable Entities for
AI Strict Liability Offenses · 112

4.3.1. AI Entity Liability · 112

4.3.2. Human Liability: Perpetration-through-Another · 114

4.3.3. Joint Human and AI Entity Liability:
Probable Consequence Liability · 116

Closing the Opening Example:
Strict Liable Killing Robot · 117

5

APPLICABILITY OF GENERAL DEFENSES TO AI CRIMINAL LIABILITY · 120

5.1. The Function of General Defenses in Criminal Law · 121

5.2. Exemptions · 123

5.2.1. Infancy · 124

5.2.2. Loss of Self-Control · 126

5.2.3. Insanity · 128

5.2.4. Intoxication · 130

5.2.5. Factual Mistake · 133

5.2.6. Legal Mistake · 136

5.2.7. Substantive Immunity · 138

5.3. Justifications · 140

5.3.1. Self-Defense · 140

5.3.2. Necessity · 144

5.3.3. Duress · 147

5.3.4. Superior Orders · 149

5.3.5. *De Minimis* · 152

Closing the Opening Example:
Robot Killing in Self-Defense · 154

6

SENTENCING AI · 156

6.1. Conceptual Applicability of Criminal
Punishment of AI Entities · 157

6.1.1. AI Purpose of Sentencing:
Combining Rehabilitation and Incapacitation · 157

6.1.2. The Legal Technique of Conversion:
The Case of Corporations (Round 2) · 162

6.2. Applicability of Punishments to AI Entities · 165

6.2.1. Capital Punishment · 000

6.2.2. Imprisonment and Suspended Imprisonment · 167

6.2.3. Probation · 169

6.2.4. Public Service · 171

6.2.5. Fine · 173

Closing the Opening Example:
Sentencing the Killing Robot · 175

Conclusions · 177

Notes · 179

Selected Bibliography · 213

Index · 241

TABLES

1.1. Target Questions of the Factual Element Components · 31

1.2. Schematic Comparison of the Requirements of Mental Element Forms · 33

1.3. Schematic Comparison of the Requirement for Satisfying the Mental Element Forms · 33

2.1. General Structure of *Mens Rea* · 49

3.1. General Structure of Negligence · 89

4.1. General Structure of Strict Liability · 109

4.2. The Legal State Reflected in the Criminal Mental State of the Other Person in Perpetration-through-Another · 115

PREFACE

In 1981, a thirty-seven-year-old Japanese employee in a motorcycle factory was killed by an artificial intelligence robot working near him. The robot erroneously identified the employee as a threat to its mission, and calculated that the most efficient way to eliminate the threat was to push the worker into an adjacent machine. Usingits very powerful hydraulic arm, the robot smashed the surprised worker into the operating machine, killing him instantly, after which it resumed its duties without further interference. This is not science fiction, and the legal question is this: Who is to be held criminally liable for this homicide?[1]

In most developed countries, unmanned vehicles, surgical robots, industrial robots, trading algorithms, personal robots, and other artificial intelligence (AI) entities are in common use. Such use may be personal, medical, military, commercial, or industrial. The question of criminal liability arises when the unmanned vehicle is involved in car accidents, the surgical robot is involved in surgical errors, the trading algorithm is involved in fraud, and so on. Who is to be held criminally liable for these offenses: the manufacturer, the programmer, the user, or the AI entity itself?

The technological world is changing rapidly. Robots and computers are more frequently replacing humans in performing simple activities. As long as humanity used computers as mere tools, there was no significant difference between computers and screwdrivers, cars, or telephones. But as computers became increasingly sophisticated, we started saying that they "think" for us. Problems began when computers evolved from "thinking machines" (devices programmed to perform specific thought processes, such as computing) into thinking machines without the quotation marks—in other words, artificially intelligent. *Artificial intelligence* is the ability of a machine to imitate intelligent behavior.

Artificial intelligence, then, is the simulation of human behavior and cognitive processes on a computer. As such, it is also the study of the nature of the entire domain of intelligent minds. Artificial intelligence research began in the 1940s and early 1950s. Since then, AI entities have become an integral part of modern human life, functioning in a much more sophisticated way than other common tools. Could these entities become dangerous? Indeed, they already are, as the incident I recounted earlier attests.

Artificial intelligence and robotics are making their way into everyday modern life. And some of the questions I have listed are currently before the courts or regulators. But there is little comprehensive analysis in the literature about assessing liability for robots, machines, or software that exercise varying degrees of autonomy.

The objective of this book is to develop a comprehensive, general, and legally sophisticated theory of the criminal liability for artificial intelligence and robotics. In addition to the AI entity itself, the theory covers the manufacturer, the programmer, the user, and all other entities involved. Identifying and selecting analogies from existing principles of criminal law, the theory proposes specific ways of thinking through criminal liability for a diverse array of autonomous technologies in a diverse set of reasonable circumstances.

Chapter 1 contains an investigation of some basic concepts on which the development of subsequent chapters is built. I will address two main questions:

1. What can be considered as an artificial intelligence entity?
2. What are the necessary requirements for the imposition of criminal liability?

Because criminal law does not depend on moral accountability, the debate about the moral accountability of machines is irrelevant to these questions. Although at times there is an overlap between criminal liability, certain types of moral accountability, and certain types of ethics, such coincidence is not necessary in order to impose criminal liability.

The question of the applicability of criminal law to machines involves two interrelated, secondary questions:

1. Is criminal liability applicable to machines?
2. Is criminal punishment applicable to machines?

The first question is a core issue of criminal law, and imposition of criminal liability depends on meeting its requirements. Only if these requirements are met does the question of punishability arise, that is, how human society can punish machines. The answers proposed in this book are affirmative for both questions. Chapters 2 through 5 examine the first question, and chapter 6 examines the second. If the affirmative answers offered in this book are accepted, then a new social entity has indeed been created, *machina sapiens criminalis*. The emergence of *machina sapiens criminalis* is examined in chapter 1. Finally, this study answers the question of whether artificial intelligence entities could ever be considered legally, if not morally, liable for the commission of criminal offenses, and deserving of punishment.

This book focuses on the criminal liability of AI entities and does not formally venture into the related areas of ethics (including roboethics) and morality. Nevertheless, the discussion of criminal liability presented here sets the stage for an in-depth consideration of the ethical issues involved, which will be necessary given the proliferation of AI entities in everyday human life. This book constructs a suitable framework for concrete and functional discussion of these issues.

I wish to thank Dr. Phyllis D. Deutsch of University Press of New England for her faith in this project from its very beginning, for guiding the publication of the book from inception to conclusion, and for her useful comments. I also wish to thank Gabriel Lanyi for his comments, insights, and disputations. I thank the anonymous readers for their comments. I also thank Ono Academic College and Northeastern University for their generous support of this project. Finally, I wish to thank my wife and daughters for their staunch support all along the way.

—G. H.

WHEN ROBOTS KILL

1

THE EMERGENCE OF *MACHINA SAPIENS CRIMINALIS*

1.1. THE ENDLESS QUEST FOR *MACHINA SAPIENS*

1.1.1. History and Prehistory of Artificial Intelligence

Since the dawn of humanity, we have sought tools to make life easier. In the Stone Age, tools were made of stone. As humans discovered the advantages of metals, metal replaced stone. With the expansion of knowledge, tools proliferated and played increasing roles in daily human life. Tools were continually challenged by increasingly complicated tasks. When they failed, more sophisticated tools were invented; if these succeeded, new challenges were posed for them, in a perpetual cycle that continues to this day. Mechanical devices have been used to ease daily life since antiquity. Heron of Alexandria, in the first century AD, designed the first steam engine and a wind-powered organ, among his many other inventions.[1] But machines became commonly used in mass production during the first industrial revolution, in the eighteenth century.

The idea of thinking machines evolved together with the insight into the human ability to create systematic methods of rational thinking. Descartes initiated the human quest for such methods in 1637,[2] although he did not believe that mechanical devices could achieve reason.[3] But Descartes laid the groundwork for the symbol-processing machines of the modern age. In 1651, Hobbes described reason as symbolic calculation.[4] During the seventeenth century, Leibniz held forth the hope of discovering a universal language of mathematics, science, and metaphysics, *characteristica universalis*, which would enable replacing thinking with calculation, and Pascal designed adding and multiplying machines, most likely the first mechanical computers,[5] operated entirely by humans and incapable of thinking. Naturally, these machines were not expected to think independently.

The modern idea of a thinking machine is generally associated with Ada Lovelace, the daughter of Lord Byron and patroness of Charles Babbage, inventor and designer of the analytical engine—the first mechanical, programmable computer, but never completed. In 1843, when she was exposed to Babbage's work, Lady Lovelace asked Babbage whether his machine could actually "think." The notion of mechanical thinking was extremely odd in those days, but nevertheless, the question itself opened the mind for considering the feasibility of unnatural, or "artificial" intelligence.

But the idea of artificial intelligence could not be examined in practice until electricity, and eventually electronic computers, became reality. In the 1950s, momentous developments in machine-to-machine translation made it possible for computers to communicate with each other, and further developments in human-to-machine translation made it possible for human operators to communicate with computers, initially in a limited fashion. In time, computer scientists learned to analyze reasoning and represent knowledge using symbols to advance the important area of artificial intelligence known as *knowledge representation*.[6]

The ability of electronic computers to store large amounts of information and process it at high speeds challenged scientists to build systems that could exhibit human capabilities. Since the 1950s, human skills and capabilities have increasingly been captured and carried out by electronic machines. With the development and popularization of the personal computer, these capabilities have become affordable for many people. Continually increasing memory capacities, speed, reliability, and robustness of personal computers are today bringing artificially intelligent tools to desktops worldwide.

During the rapid advances in the field of computers in the 1950s, artificial intelligence (AI) developed as a separate area of research that combined technological investigation with the study of logic and eventually of cybernetics, an interdisciplinary field dealing with communication between humans and machines, among other issues. Studies in logic in the 1920s and 1930s helped formalize the methods of reasoning, and produced a new form of logic known as propositional and predicate calculus, relying on the works of Church, Gödel, Post, Russell, Tarski, Whitehead, Kleene, and others.[7] Developments in psychology, neurology, statistics, and mathematics in the 1950s were also incorporated into the growing research sphere of AI.[8]

In the late 1950s, several developments took place that signaled the emergence of AI to the public at large. The most prominent of these was the development of chess-playing programs and of the General Problem Solver (CPR), designed to solve a wide range of problems, from symbolic integration to word puzzles. This brought the abilities of AI to the attention of an enthusiastic public, creating at the same time unrealistic expectations and also the first science fiction novels about robots rebelling against humans and establishing control over them.[9]

AI research proceeded in two main directions: (1) building physical devices combined with digital computers, and (2) developing symbolic representations. The first effort led to robotics; the second sought to train perception to classify certain types of patterns as either *similar* or *distinct*. AI research was guided by the assumption that the problem of commonsense knowledge is solvable, and that if humans can solve it, so can machines. The problem was formulated as the ability to glean facts in situations when not all the information is given, using common sense. For many years, this assumption prevented progress in theoretical AI and diverted AI research toward connectionism (neural network models).[10]

By the 1970s, the importance of AI became apparent to most of the world. Governments in developed and developing countries were seeking long-term commitments of resources to fund intensive research programs in AI.[11] Government and private organizations routinely cooperated in development programs in the areas of robotics, software, and various computer products. These projects were driven by the realization that it had become feasible to develop systems that could exhibit such human abilities as understanding speech and visual scenes, learning and refining knowledge, and making independent decisions.

Beginning in the 1970s, industry embraced AI technology. Some AI capabilities, such as natural language processing, have not yet been fully implemented, but in a growing number of fields, AI technology is being applied successfully by industry.[12] Examples of such areas are biomedical microscopy, materials analysis, and robotics. Areas in which AI has been successfully applied have become entirely dependent on this technology because common human abilities cannot replace the accuracy, speed, and efficiency of these AI tools.

In the 1980s, AI research made giant gains in the design and development of expert systems in the fields of medicine, finance, and anthropology. The main challenge of expert systems was to develop suitable knowledge representations in their respective fields. To be readily accessible, this knowledge needed to be stored in a form that could be retrieved and displayed automatically by the system interfaces, human or other. Many expert systems became successful tools, expanded over the years with new knowledge and improved with better heuristics.[13]

The next challenge was to enable new technologies to be incorporated into expert systems shortly after they became available. Development of expert systems resulted in intensive research in the areas of machine learning and problem solving, which expanded the use of expert systems into many new fields, with development efforts extending beyond the confines of academic research, and AI technology increasingly replacing traditionally human capabilities.[14] With the growing need for AI development in industry, the advantages of AI technology were no longer subjects of debate.

The involvement of industry in AI research grew for three main reasons. First, the achievements of AI, especially in knowledge engineering, were beyond any doubt. Second, advances in hardware made the development of AI tools more feasible and the products more accessible to users. Third, industry had a growing need for faster and better problem-solving tools in order to increase productivity. Industry, therefore, adopted the use of AI technology in many areas, such as factory automation, programming, office automation, and personal computing.[15]

The combination of improving abilities in AI technology, of human curiosity, and of industry needs is responsible for the global expansion in the use of AI technology. This trend continues into the third millennium, as more traditionally human functions are being replaced by AI technologies. For example, the South Korean government now uses AI robots as soldiers to guard the border with North Korea; as teachers in schools; and since 2012, as prison guards.[16]

In an eighty-two-page report that outlines the future use of drone aircraft, "Unmanned Aircraft Systems Flight Plan 2009–2047," the US Air Force writes that autonomous drone aircraft are key "to increasing effects while potentially reducing cost, forward footing, and risk." The report states that, similar to a chess program that can outperform proficient chess masters today, future drones will be able to react faster than human pilots ever could. But the report also notes potential legal problems: "Increasingly humans will no longer be 'in the loop' but rather 'on the loop'—monitoring the execution of certain decisions . . . Authorizing a machine to make lethal combat decisions is contingent upon political and military leaders resolving legal and ethical questions."[17]

1.1.2. Defining Artificial Intelligence, and the Endlessness of the Quest for *Machina Sapiens*

Since the beginning of AI, researchers have been trying to develop computers that actually "think." This is the holy grail of AI research.[18] But to develop a thinking machine, it is necessary to first define what exactly *thinking* is. Defining thinking, with regard to both humans and machines, proved to be a complicated task for AI researchers. The development of machines that have the ability to think independently would be an important event for humankind, which has claimed a monopoly over this high mental skill. The creation of a true thinking machine would be tantamount to the emergence of a new species on earth, the *machina sapiens*.

But does human science want to create a new species? Since time immemorial, people have been trying to play God, with the first preventing steps mentioned in the Bible.[19] Some aspects of modern scientific research contain elements of similar conceit, including such endeavors as human cloning, biomedical engineering, anti-aging pursuits, and more. The quest for the creation of a new species matches this trend. Although creating a new species may benefit humans, this is not necessarily the motivation behind AI research. The reason may be much deeper, touching on the most profound of human quests, which, according to the Bible, was prohibited to humans following the original sin.

The first move toward the creation of *machina sapiens* is to define artificial intelligence. Various definitions have been proffered. Bellman defined it as "the automation of activities that we associate with human thinking, activities such as decision-making, problem solving, learning."[20] Haugeland defined it as "the exciting new effort to make computers think . . . machines with mind, in the full and literal sense."[21] Charniak and McDermott defined it as "the study of mental faculties through the use of computational models."[22] Schalkoff defined it as "a field of study that seeks to explain and emulate intelligent behavior in terms of computational processes."[23] Kurzweil defined it as "the art of creating machines that perform functions that require intelligence when performed by people."[24] Winston defined it as "the study of the computations that make it possible to perceive, reason, and act."[25] Luger and Stubblefield defined it as "the branch of computer science that is concerned with the automation of intelligent behavior."[26] Rich and Knight defined it as "the study of how to make computers do things at which, at the moment, people are better."[27]

At first sight, these definitions seem more confusing than clarifying. But according to these and other definitions, AI systems may be grouped into four main categories: those that (1) act like humans, (2) think like humans, (3) think rationally, and (4) act rationally.[28]

AI systems that act like humans are validated by the *Turing test*, proposed by the British mathematician in 1950 to provide an operational definition of intelligence.[29] Turing defined intelligent behavior as the ability to achieve human-level performance in all cognitive tasks that would prevent a human interrogator from distinguishing between human and machine behavior. The Turing test involves a human "listening" to a conversation between a machine and a human. The conversation is carried out in writing to avoid identifying the machine through its voice; therefore, the "listening" is carried out in reading the text of the questions and answers. The machine is considered to have passed the Turing test if the human interrogator cannot tell the difference between the two.[30] The Turing test assumes equal cognitive abilities for all humans, although conversations between a machine and a child, a person with an intellectual disability, a very tired person, and the designer of the machine are likely to be significantly different.[31]

To identify AI systems that think like humans, it would be necessary to define human thinking first. AI technologies developed as general problem solvers were designed to make decisions that are very similar to human decisions based on the same information.[32] The modern development of cognitive science enabled experimental approaches to mechanical thinking, and to the testing of machine thinking. In another test proposed by Turing, the interrogator's objective is to differentiate between a man and a woman in a conversation, a test that depends on the communicative skills of the human participants no less than on that of the machine.[33]

The effectiveness of the original Turing test has been questioned, especially regarding *strong* AI (in which computers have a mind similar to that of humans). Searle, in his famous Chinese Room thought experiment,[34] described a person who does not know Chinese, alone in a room, receiving batches of Chinese messages in writing. A rule book, written in the person's native language, allows him or her to look up the Chinese characters by their shape, and then to answer the messages using this rule book. People outside the room, who are conducting the Turing test, are convinced that the person inside the room understands Chinese.

In this thought experiment, the person inside the room simply follows the instructions in the book, but neither the person nor the book understands Chinese, although both can simulate such understanding. The instruction book is, in this case, a computer program. What, then, does it mean for humans to understand a foreign language? It is difficult to identify an AI system that thinks rationally without defining *rationality* first. If rationality means "right" thinking, it can be represented by formal logic, so that given correct information, the machine can aim to produce a combination of right conclusions. For example, knowing that all tomatoes are red and that X is a tomato, the machine should reach the conclusion that X is red. Most modern AI systems support formal logic and know how to behave in accordance with it.

AI systems that act rationally are an advanced variation of systems that think rationally. The latter can reach the right conclusions from correct information as outsiders, but the former can participate in events, taking the right action based on correct information. For example, an AI goalkeeper in a soccer game that sees the ball approaching the goal rapidly can not only calculate the speed and angle at which the ball is approaching and determine the action required to block it, but also take the required action and stop the ball.

The quest for *machina sapiens*, however, reaches much deeper than these classification and definitions, and if the quest is successful, it will answer the question of what makes humans intelligent, and will result in the design of intelligent machines. Since the late 1980s, the accepted approach has been that intelligent thinking is identified by certain attributes, specifically the following five: (1) communication, (2) internal knowledge, (3) external knowledge, (4) goal-driven conduct, and (5) creativity.[35] Let us discuss each of these attributes.

Communication is considered the most important attribute that defines an intelligent entity, because we can communicate with intelligent entities. Note that we communicate not only with humans, but with certain animals as well, although the range of such communication is narrower than our range with humans, and not all ideas can be expressed. For example, you can let your dog know how angry you are, but you cannot explain to him the Chinese Room thought experiment. This situation is not different from communicating with a two-year-old human. The more intelligent the other entity, the more complex the communication conducted with it.

Communication assumes proper understanding of the information contained in it, and can be used to test the ability to understand complicated ideas. But communication does not always attest to the quality of understanding. Some highly intelligent people, at genius level, are not good communicators, and some autistic geniuses cannot communicate at all. At the same time, most normative people have advanced communication skills, although only a few can communicate about complex matters such as Einstein's general theory of relativity, quantum mechanics, parallel universes, or exponential equations.

Communication can be carried out not only through speech, but by other means as well, including writing. Therefore, machines can be quite intelligent even if they lack the ability to speak, similar to mute people. So although the communication attribute of intelligence is important, there are many exceptions to it. The Turing test is based on this attribute, but if human communication is so inaccurate, how can we trust it to identify AI?

Internal knowledge refers to the knowledge of entities about themselves. This parallels self-awareness. An intelligent entity is supposed to know about its own existence, about the fact that it functions in some way, about the way in which it integrates into factual reality, and so on. Formal logic reasoning showed the way to artificial internal knowledge through self-reference.[36] It is possible to program computers to appear that they know about themselves, and that they know about knowing about themselves. But many researchers consider this too artificial, and insist that it would be difficult to determine whether these computers *really* know about themselves. Yet no alternatives have been suggested to test internal knowledge. The question is, how is it possible to know anything about another person's internal knowledge?

External knowledge refers to factual data about the outside world and about factual reality. This attribute is considered crucial in an age when knowledge functions as a commodity, especially with relation to expert systems.[37] An intelligent entity is expected to know how to find data about the outside world, and to know the facts that make up the factual reality to which it is exposed. This attribute assumes memory skills and the ability to classify information into seemingly relevant categories. This is the way humans assemble their life experience and the way in which they learn. It is difficult to act as an intelligent entity if all factual elements are treated each time as new. Although factual events are new each time, they do contain common characteristics that an intelligent entity should identify.

For example, medical expert systems, which are designed to diagnose diseases based on their symptoms, should be able to identify the common characteristics of diseases in many cases, even if these cases vary to an extreme degree. An entity that lacks this ability acts in a way similar to people who suffer from amnesia. Although they conduct themselves adequately in a given circumstance, their conduct is not added to their cumulative experience. This is how simple machines operate: performing a certain task, but not knowing that they have done so, and not having the ability to draw on this or other experiences to guide them in future tasks. So the question is, if inexperienced people can still be considered intelligent, then why not machines that behave similarly?

Goal-driven conduct marks the difference between random or arbitrary conduct and intended conduct. Goal-driven conduct requires an operative plan to achieve the desired goal. For most humans, goal-driven conduct is interpreted as intention. If one is hungry and sees an apple, eating it is goal-driven conduct, the goal being to appease the hunger. We may say that the person ate the apple with the intention of not being hungry anymore. Goal-driven conduct is not unique to humans. When a dog sees a juicy bone behind an obstacle, she plans to bypass the obstacle to get the bone. Executing the plan is goal-driven conduct on the part of the dog.

Different creatures may have different goals, of differing levels of complexity. The more intelligent the entity, the more complex are its goals. The goals of dogs are at the level of calling for help when their masters are in distress, whereas the goals of humans may include landing on Mars. Computers have the ability to plan the landing on Mars, and robots and unmanned vehicles are already exploring Mars, searching for data. The reductionist approach to goal-driven conduct deconstructs a complex goal into many simple ones; achieving these is considered goal-driven conduct. Goals and plans for achieving them may be incorporated into computer programs. But not all humans pursue complex goals at all times and under all circumstances. The question is, what level of complexity of goals is needed for a machine to be considered intelligent?

Creativity involves finding new ways of understanding or doing things. All intelligent entities are assumed to have some degree of creativity. When a fly tries to get out of the room through a closed window, it crashes against the windowpane over and over again, trying invariably the same conduct without success. This is the opposite of creativity. At some point, however, the fly does seek another way out, which is considered to be more creative. A dog may find a way out in much fewer attempts; a human in fewer still. Computers may be programmed not to repeat the same conduct more than once, and instead to seek other ways of solving the problem. This type of behavior is essential for general problem-solving software.

Creativity is not homogenous, and has degrees and levels. Not all humans are considered to be thinking outside the box, and many perform their daily tasks exactly the same way day after day. Most lottery players play the same numbers time after time for years without winning. What makes their creativity different from that of the fly? The fly's chance of breaking through the window is not smaller than that of the lottery player winning the main prize. Many factory workers perform the same operations for hours every day, and they are considered intelligent entities. The question is, what is the exact level of creativity required to be considered intelligent?

Not all humans who are considered intelligent share all five attributes. Why, then, should we use different standards for humans and machines to measure intelligence? But each time some new software has succeeded in conquering a given attribute, the achievement was rejected for not being *real* communication, internal knowledge, external knowledge, goal-driven conduct, or creativity.[38] Consequently, new tests were proposed to make sure that the relevant AI technology was *really* intelligent.[39]

Some of these tests have had to do with knowledge representation (what the machine knows), decoding (translating knowledge from factual reality to its representation), inference (extracting the content of knowledge), control of combinatorial explosion (preventing endless calculation for the same problem), indexing (arranging and classifying knowledge), prediction (assessing the probability of possible factual events), dynamic modification (self-changing the programs as a result of experience), generalization (inductive interpretation of factual events), and curiosity (wondering about factual events or seeking reasons for them).

In their biological sense, all the attributes just mentioned are manifestations of the human brain. Nobody doubts this fact. These attributes are manifestations of neuronal activity in the brain, which can be identified and quantified. If so, and if transistors can be functionally activated in the same way as neurons, why can these attributes not be reproduced by means of transistors?[40] The simple answer is that AI entities are simply not human.

The process of proposing new tests whenever a given AI technology succeeds in passing tests devised earlier makes the quest for *machina sapiens* endless. The reason for this is psychological rather than technological. When talking about AI, we must not envisage regular human beings with a robotic-metallic appearance. Yet we will settle for nothing less,[41] forgetting that artificial intelligence happens to be *artificial*, not natural. When AI technology finally succeeds in passing a certain test, this shows us that the problem was not in the technology, but in the test. Because the complexity of the human mind is too vast to be tested with simple tests, we replace the tests, a process in which we have learned more about the human mind, especially about the "bureaucracy" of mind and its intentionality,[42] than about technology.[43]

The template of the arguments against the definition and feasibility of *machina sapiens* as a machine that possesses truly intelligent personhood goes like this:

1. To possess personhood, an entity must have attribute X.
2. AI technology cannot possess attribute X.
3. Behavior of AI technology that is identified to possess attribute X demonstrates that it can simulate or imitate that attribute.
4. Simulation of attribute X is not attribute X itself.

Therefore, AI technology is not really intelligent. Some scholars called this template of arguments a "hollow shell strategy."[44] Note that argument number 2 and the conclusion result in a catch-22: AI entities are not intelligent because AI technology cannot possess the attribute of intelligence.

The contents of attribute X may be made up entirely of advanced intelligence tests, which in this context reflects the belief that intelligence can be only human, characterized by the fact that it is exclusively human, so that machines cannot perform it. The result is a paradox. Despite remarkable advances in AI technology, there has been great frustration with the abilities of AI, as progress in AI research has underscored how far the technology is from imitating the human mind and how much farther the mysteries of the human mind still need to be explored.

In the 1950s, it was difficult to believe that a computer could ever defeat humans in chess, and that if it were to happen, computers would be considered truly "intelligent." But in 1997 a computer program (Deep Blue) defeated the world champion in chess, and it was still not considered intelligent.[45] In 2011 another computer program (Watson) defeated two of the top *Jeopardy* players on television, but it was still not considered intelligent.[46] Although the computer missed some of the jokes on the show, it won the contest. The technology developed for Watson is being adapted for use in advanced expert systems, and although it is considered good enough to diagnose diseases, many people still do not consider it intelligent.

AI technologies pioneered *machine learning*, an inductive method of learning in which the computer analyzes various cases and situations submitted to it, and by means of generalizations, produces an image of the facts for future use.[47] Similar knowledge in humans is called *expertise*, but humans prefer not to refer to these computer systems as intelligent. It seems that the creation of *machina sapiens* recedes further with each advance in AI technology, and that AI research has produced an endless quest for the new species.

There are two ways of dealing constructively and positively with the endless quest for *machina sapiens*. The first is through technological research, seeking ways to reduce the gap between machines and humans. This is the path chosen by AI researchers since 1950, following the original quest for *machina sapiens* in the belief that some day technology will be able to imitate the human mind. AI research has had significant accomplishments by following this method.[48] The second way is industrial. As I will describe, industry is not interested in imitating the human mind in all its details. Indeed, industry welcomes the opportunity of using AI entities that do *not* exhibit certain human qualities. So the disadvantages of machines—the limited ways in which they imitate the human mind—become advantages for industry. Turning the disadvantages of the machine into advantages for industry makes it possible to increase the use and integration of AI entities in industry. The industrial use of AI technology was the catalyst for the emergence of *machina sapiens criminalis*.

1.1.3. Turning Disadvantages into Advantages: Industrial and Private Use of Artificial Intelligence

Industrial use of artificial intelligence technology is not new. As noted earlier,[49] AI technology has been embraced in advanced industries since the 1970s. But whereas in the beginning, AI technology was used by industry because of its similarity to the human mind, later it was used because of its *differences* from the human mind. Industry was quick to understand that complete and perfect imitation of the human mind would not be as useful as incomplete imitation, so industry encouraged the development of AI technology as long as imitation of the human mind was not complete. And because complete imitation of the human mind is still far in the future, industry and AI research continue to cooperate.

How did industry turn a disadvantage into an advantage? Consider the calculator. If we enter "$1 \times 1 =$" into a calculator repeatedly, we continue to receive the answer "1" invariably, even if we repeat this operation thousands of times. The process activated by the calculator is the same each time. But if we ask a human the same question, we may receive an answer the first time — unless the human believes she is being teased — and perhaps several more times, but not thousands of times. At some point, the human stops answering because she becomes bored, irritated, or nervous, or because she loses any desire to answer.

This is a major problem from the point of view of AI researchers, and it underscores the fact that the human mind may act arbitrarily, sometimes for irrational reasons. It would be unthinkable, however, for the calculator to refuse to answer "$1 \times 1 =$" even for the thousandth time. For tasks of this type, industry prefers entities that are not bored with our requests or whims, are not irritated by our questions, and will continue to serve us well, no matter how many times we ask the same question.

It appears that most humans lack these abilities because they have a human mind. But machines, which have not succeeded in imitating the human mind completely, have the ability to provide us with this kind of service. The example just presented is theoretical, of course, as no one would really type a thousand times "$1 \times 1 =$" on one's calculator, a task that in itself would require nonhuman skills. But machine skills of this nature are required in most industries.

Consider the customer service representatives of large companies that serve hundred of thousands of customers and must provide polite, helpful answers to requests of every type. How do these representatives respond after one call? After a hundred calls? After thousands of calls? How does the number of calls affect the quality of service? But the technological inability of a machine to become bored, irritated, or tired is a net advantage for industry. An automatic customer-service system services the thousandth customer exactly the way it serviced the first one: politely, patiently, efficiently, and accurately.

Expert systems specializing in medical diagnosis are not bored by repeating identical solutions to problems of different patients. Police robots are not too frightened to dismantle dangerous explosive devices. Factory robots are not bored by performing identical activities thousands of times a day. Industry has leveraged the nonhuman qualities of AI entities, turning the traditional disadvantages of AI technology (at least from the point of view of AI researchers) into advantages for modern industry.[50]

In practice, these disadvantages are benefitting not only industry. Research in AI technology and industry has made this technology available for private consumption as well. Personal robot assistants based on AI technology are already available, and AI robots are expected to make their appearance in family and private life, including some of the most intimate situations. It has already been suggested that "love and sex with robots" may provide a better and socially healthier alternative to prostitution.[51] No shame, abuse, or mental or physical harm would result from using the AI alternative, and this could result in a real social change.

Household robots are not offended if they are asked to perform the same action over and over, and they do not require vacations, demand raises, or ask for favors. Teacher robots are unlikely to teach material outside the curriculum they have been programmed to teach, and prison-guard robots are unlikely to be bribed to allow prisoners to escape. These nonhuman qualities have made AI technology highly popular for meeting industrial and private needs.

We can characterize the AI technology required for industrial and private needs as a less than complete and perfect imitation of the human mind, but as having some human skills and being capable of an imperfect and incomplete imitation of the human mind. The resulting AI entities are not yet *machina sapiens*, but they possess some human skills for solving problems and can imitate some of the capabilities of the human mind. The existing capabilities of AI technology, which are already being used to satisfy industrial and private needs, are incorporated today in the *machina sapiens criminalis*.

1.2. ON EVOLUTION AND DEVOLUTION: *MACHINA SAPIENS CRIMINALIS*

In modern times, technology and law often find themselves on opposite sides of the spectrum. Whereas technology serves the cause of the evolution of new innovations, the law pulls in the conservative direction. For example, technology made human cloning possible in principle, yet in most countries, the law prohibited it or initiated a public debate about it. What technology considers progress, the law may treat as regress. At times the law is inspired by moral conceptions of society, at other times by human fears. This is the situation with advanced robotics based on AI technology that has produced the *machina sapiens criminalis*.

1.2.1. Serve and Protect: Human Fears of Human-Robot Coexistence

Since the beginning of the twenty-first century, reports about research in advanced technology have predicted that starting at about the third or fourth decade of this century, a *new generation of robots* will support human-robot coexistence.[52] Under the Fukuma "World Robot Declaration" issued in 2004, these robots are anticipated to coexist with humans, assist humans both physically and psychologically, and contribute to the realization of a safe and peaceful society.[53]

It has been accepted that there are two major types of such robots. The first is a new generation of industrial robots capable of manufacturing a wide range of products, performing multiple tasks, and working with human employees. The second is a new generation of service robots capable of performing such tasks as house cleaning, security, nursing, life support, and entertainment by coexisting with humans in home and business environments. The authors of most published projections on this technology have also evaluated the level of danger to humans and society as a result of using these robots.[54] Such evaluations provoked a debate about the safety of using robots, irrespective of the real level of predicted danger.

People consider safety only when something is thought dangerous. The accelerated technological developments in the area of AI and robotics caused many fears. For example, one of the first reactions to a humanoid robot as a caregiver designed to provide nursing care was the fear of hurting the assisted human. Would all humans be ready to let their babies and children be nursed by robots? Most humans are not experts in technology, and most fear what they do not know.

Further, the vacuum created by the absence of knowledge and certainty is sometimes filled by science fiction. In the past, science fiction was rare

THE EMERGENCE OF *MACHINA SAPIENS CRIMINALIS* · 15

and had a small following. Today, Hollywood brings science fiction to people worldwide. Most blockbusters of the 1980s, 1990s, and twenty-first century are classified as science fiction movies, and most of these films exploit people's fears. For example, in *2001: A Space Odyssey* (1968), based on Arthur C. Clarke's novel, the central computer of the spaceship is out of human control and attempts to assassinate the crew.[55] Safety is restored, after many deaths, only when the computer is shut down.[56]

In *The Terminator* series of films (*The Terminator*, 1984; *Terminator 2: Judgment Day*, 1991; and *Terminator 3: Rise of the Machines*, 2003), machines are taking over humanity, which is almost extinct. A few survivors establish a resistance to oppose the machines, which must be shut down to ensure human survival. Even the savior, which happens to be a machine, ends up shutting itself down.[57] In the trilogy of *The Matrix* (*The Matrix*, 1999; *Matrix Reloaded*, 2003; and *Matrix Revolutions*, 2003), machines dominate the earth, enslaving humans, to produce energy for their own benefit. The machines control the humans' minds by creating the illusion of a fictional reality called the "matrix." Only a few can escape the matrix to fight for freedom from the domination of the machines.[58] In *I, Robot* (2004), based on an Isaac Asimov collection of stories from 1950, an advanced model of robots hurts people, one robot is suspected of murder, and one of the human heroes is a detective who does not trust robots. The overall plot concerns the robots' attempt to take over humans.[59]

The influence of Hollywood is vast, so it is reasonable to expect that fear of artificial intelligence and robotics will dominate the public mind. The more advanced a robot, the more dangerous. A popular topic in science fiction literature and films is revolt and takeover by robots. Thus, when people think about advanced robots and technology, in addition to utility and unquestionable advantages, they also consider how to protect themselves. People will accept the idea of human-robot coexistence only if they feel safe from the other, the different, the terrifying: the machine.

What protection mechanisms can humanity use to ensure the safety of coexistence with robots and AI technology?[60] The first to sound the alarm about the dangers of such technology was science fiction literature, which was also the first to suggest protection from it. The first circle of protection to be proposed was in the realm of ethics focused on safety. The ethical issues were presented to the designers and programmers of AI entities, to ensure that built-in software would prevent any unsafe activity of the technology.[61] One of the pioneers of the attempts to create robot ethics was Isaac Asimov.

Asimov stated three famous "laws" of robotics in his collection of nine science fiction stories assembled under the title *I, Robot* in 1950: "(1) A robot may not injure a human being or, through inaction, allow a human being to come to harm; (2) A robot must obey the orders given it by human beings, except where such orders would conflict with the First Law; (3) A robot must protect its own existence, as long as such protection does not conflict with the First or Second Laws." [62] In the 1950s, these laws were considered innovative, and restored some confidence in a terrified public. After all, harming humans was not allowed.

Asimov's first two laws represent a human-centered approach to safety in relation to AI robots. They also reflect the general approach that as robots gradually take on more intensive and repetitive jobs outside industrial factories, it is increasingly important that safety rules support the concept of human superiority over robots.[63] The third law straddles the border between human-centered and machine-centered approaches to safety. The functional purpose of the robots is to satisfy human needs, and therefore robots should protect themselves in order to better perform these tasks, functioning as human property.

But these ethical rules, or laws, were insufficient, ambiguous, and not sufficiently broad, as Asimov himself admitted.[64] For example, a robot in police service trying to protect hostages taken by a criminal who intends to shoot one of them understands that the only way to stop the murder of an innocent person is to shoot the criminal. But the first law prohibits the robot from killing or injuring the criminal (by acting), and at the same time prohibits it from letting the criminal kill the hostage (through inaction). What can we expect the robot to do based on the first law, with no other alternatives available? Indeed, what would we expect a human police officer to do? Any solution breaches the first law.

Examining the other two laws does not change the consequences. If a human commander ordered the robot to shoot the criminal, the robot would still be breaching the first law. Even if the commander were in immediate danger, it would be impossible for the robot to act. And even if the criminal threatened to blow up fifty hostages, the robot would not be allowed to protect their lives by injuring the criminal. Matters would be even more complicated if the commander were a robot too, but the consequences would be no different. And the third law would not alter the results even if the robot itself were in danger.

The dilemma of the police robot is not uncommon in a society in which humans and robots coexist. Any activity by robots and AI technology in such a society is riddled with such dilemmas. Consider a robot in medical service that is required to perform an intrusive emergency surgical procedure intended to save a patient's life, and to which the patient objects. Any action or inaction by the robot breaches Asimov's first law, and any order by a superior is unable to solve the dilemma because an order to act causes injury to the patient and an order to refrain from action results in the patient's death.

Some dilemmas are easier to solve when one of the options involves no injury to humans, but this raises other questions. In the previous example of robots used as prison guards,[65] what should such a robot do when a prisoner attempts to escape and the only way to stop the prisoner is by causing injury? What should a sex robot do when ordered to commit sadistic acts? If the answers are *not to act*, then the question is, why use such robots in first place if their intended actions breach the first law?

If humanity determines that the destiny of AI robots and other artificial entities is to serve and protect humanity in various situations, this involves difficult decisions that will have to be made by these entities. The terms *injury* and *harm* may be wider than specific bodily harm, and these entities may harm people in other ways than bodily. Moreover, in various situations, causing one sort of harm should be necessary in order to prevent greater harm. In most cases, such decision involves complicated judgment that exceeds the simple dogmatic rules of ethics.[66]

The debate about Asimov's laws gave rise to debate about the moral accountability of machines.[67] The moral accountability of robots, machines, and AI entities is part of the endless quest for *machina sapiens*. But robots exist, they participate in daily human life in industrial and private environments, and they do cause harm from time to time, regardless of whether they can be subjected to moral accountability. Therefore, the ethical sphere is unsuitable for settling the issue. Ethics involve moral accountability, complicated inner judgment, and inapplicable rules. The question, then, is what can settle robot functioning, human-robot relations, and human-robot coexistence? The answer may lie in criminal law.

1.2.2. Evolution and Devolution:
Machina Sapiens Criminalis as a By-product

Artificial intelligence technology makes possible a range of industrial and private uses of robots. The technology will certainly become more advanced in the future as AI research develops over time, and both industrial and private use of this technology will broaden the range of tasks that AI robots will undertake. The more advanced and complicated the tasks, the higher the chances of failure in accomplishing them. *Failure* is a broad term that includes various situations in this context. The most common situation is that the task undertaken by the robot has not been accomplished successfully. But some failure situations can involve harm and danger to individuals and society.

For example, the task of prison-guard robots has been defined as preventing escape by using minimal force against the prisoners. A prisoner attempting to escape may be restrained by the robot guard, which holds the prisoner firmly but causes injury; the prisoner may then argue that the robot has excessively used its power. Analyzing the robot's actions may reveal that it could have chosen a more moderate action, but the robot had evaluated the risk as being graver than it actually was. In this case, who is responsible for the injury?

This type of example raises important questions and many arguments about the responsibility of the AI entity. If analyzed through the lens of ethics, the failure in this situation is that of the programmer or the designer, as most scientists would argue, not of the robot itself. The robot cannot consolidate the necessary moral accountability to be responsible for any harm caused by its actions. According to this point of view, only humans can consolidate such moral accountability. The robot is nothing but a tool in the hands of its programmer, regardless of the quality of its software or cognitive abilities. This argument is related to the debate about *machina sapiens*.

Moral accountability is indeed a highly complex issue, not only for machines, but for humans as well. Morality, in general, has no common definition that is acceptable in all societies by all individuals. Deontological morality (concentrated on the will and conduct) and teleological morality (concentrated on the result) are the most acceptable types, and in many situations they recommend opposite actions.[68] The Nazis considered themselves deontologically moral, although most societies and individuals disagreed. If morality is so difficult to assess, then moral accountability may not be the most appropriate and efficient way of evaluating responsibility in the type of case we have just examined.

In this context, the issue of the responsibility of AI entities will always return to the debate about the conceptual ability of machines to become human-like, so that the endless quest for *machina sapiens* would become an endless quest for AI accountability. The relevant question here exceeds the technological one, and it is mostly a social question. How do we, as a human society, choose to evaluate responsibility in situations of harm and danger to individuals and society?

The main social tool available for handling such situations in daily life is criminal law, which defines the criminal liability of individuals who harm society or endanger it. Criminal law also has educational social value because it educates individuals on how to behave within their society. For example, criminal law prohibits murder; in other words, the law defines what is considered to be murder, and prohibits it. This has the value of punishing individuals for murder *ex post*, and prospectively educating individuals not to murder *ex ante*, as part of the rules of living together in society. Thus, criminal law plays a dominant role in social control.[69]

Criminal law is considered the most efficient social measure for directing individual behavior in any society. It is far from perfect, but under modern circumstances it is the most efficient measure. And because it is efficient regarding human individuals, then it is reasonable to examine whether it is efficient regarding nonhuman entities as well, specifically AI entities. Naturally, the first step in evaluating the efficiency of criminal law with regard to machines is to examine the applicability of criminal law to them. So the question in this context is whether machines may be subject to criminal law.

Because criminal law does not depend on moral accountability, the debate about the moral accountability of machines is irrelevant to the question at hand. Although at times there is an overlap between criminal liability, certain types of moral accountability, and certain types of ethics, such coincidence is not necessary in order to impose criminal liability. The question of the applicability of criminal law to machines involves two interrelated, secondary questions:

1. Is criminal liability applicable to machines?
2. Is criminal punishment applicable to machines?

The first question is a core issue of criminal law, and imposition of criminal liability depends on meeting its requirements. Only if these requirements are met does the question of punishability arise, that is, how human society can punish machines. The answers proposed in this book are affirmative for both questions. Chapters 2 through 5 examine the first question, and chapter 6 examines the second. If the affirmative answers offered in this book are accepted, then a new social entity has indeed been created, *machina sapiens criminalis*.

Machina sapiens criminalis is the inevitable by-product of human efforts to create *machina sapiens*. Technological response to the endlessness of the quest for *machina sapiens* has resulted in advanced developments in AI technology, which have made it possible to imitate human minds and their associated skills better than ever before. Current AI technology is capable of performance that was considered science fiction only a few years ago—a situation that has existed over the entire second half of the twentieth century, and continues into the twenty-first century.

Every step along the road of technological development is another step in the evolution of *machina sapiens*. The endless quest for *machina sapiens* created very high upper and lower thresholds, so high in fact that today, it is necessary to be human to be considered *machina sapiens*. So the quest remains endless, and the technological race continues. But criminal liability does not necessarily require the presence of *all* human skills. To become an offender, one need not use *all* human skills, regardless of whether one possesses such skills.

Consider the attribute of creativity, discussed earlier.[70] Naturally, there are many types of thieves. Some are deemed creative, others not. When examining their criminal liability for a given offense, their human skills of creativity are not even considered. We impose criminal liability for theft regardless of whether the offender was creative in committing the theft. Because creativity is not a condition for the imposition of criminal liability on any offender, it is not considered in the legal process.

When the requirements of criminal liability for the specific offense are met, no other qualifications, skills, or thoughts are considered. It is possible to argue that although theft does not require creativity, humans are assumed to be creative, and therefore criminal liability is applicable to them. This type of argument may be relevant in the debate on the endlessness of the quest for *machina sapiens*, but absolutely not in the question of criminal liability. Human skills that are irrelevant for the commission of a given offense are not considered as part of the legal process toward criminal liability.

Therefore, whether the offender has been creative in committing the offense is irrelevant for the imposition of criminal liability, because there is no such requirement for criminal liability. Whether the offender was or was not creative outside the commission of the offense is also irrelevant, because the legal process focuses only on the facts involved in the commission of the offense. This is also true of many other attributes considered necessary for announcing the birth of *machina sapiens*.

It appears, therefore, that in the race to create *machina sapiens*, significant by-product has been developed: a type of machine that does not have the skills to be considered *machina sapiens* because it cannot fully imitate the human mind; but its skills are adequate for various activities in industrial and private environments. Indeed, the absence of some skills is even considered an advantage when focusing on industrial and private use, although AI research considers such absence a disadvantage.[71] One type of activity that this by-product is capable of is the commission of offenses.

In other words, although the by-product is not capable of many types of creative activities, it is perfectly capable of committing offenses. The reason for this lies in the definitions and requirements of criminal law for the imposition of criminal liability and punishment. These requirements are satisfied by capabilities that are far lower than those required to create *machina sapiens*. In the context of criminal law, as long as these requirements are met, there is nothing to prevent the imposition of criminal liability, whether the subject of criminal law is human or not. This is the logic that applies to human and nonhuman offenders, such as corporations.

Thus, as long as all relevant requirements of criminal law are met, a new type of subject may be added to the large group of existing subjects to criminal law, in addition to human individuals and corporations. These subjects may be referred to as *machina sapiens criminalis*, not a subtype of a general *machina sapiens*, but its by-product. This by-product is considered to be less advanced technologically, because it is possible to belong to this category of machines without the high-level skills attributed to the ideal *machina sapiens*, if and when achieved.

From the technological-scientific point of view, any progress toward the development of *machina sapiens* is considered progress (or "evolution"),[72] whereas the by-product, *machina sapiens criminalis*, is more of a regress (or "devolution"). *Machina sapiens criminalis* is a stopping point along the race to the top. It is a less-advanced peak, because the requirements for criminal liability are much lower than those for the ideal *machina sapiens*. Nevertheless, *machina sapiens criminalis* is not a race to the bottom, but merely a stopping point in the race to the top. From the criminal law point of view, the technological research may rest at this point because the entities to which the criminal law is applicable already exist.

If the range of machines extends between the poles of technically basic ("dumb") machines and *machina sapiens*, then *machina sapiens criminalis* is situated somewhere in the middle, perhaps closer to the *machina sapiens* pole. But as noted earlier, the emergence of *machina sapiens criminalis* is contingent on a match between the requirements of criminal law and the relevant skills and abilities of the machine. Some researchers argue that the current law is inadequate for dealing with AI technology, and that it is necessary to develop a new legal domain called *Robot Law*.[73]

The argument of this book is that the current criminal law is adequate to cope with AI technology. Moreover, if technology were to advance significantly toward the creation of *machina sapiens*, this would make current criminal law even more relevant to addressing AI technology because such technology imitates the human mind, and the human mind is already subject to current criminal law. Thus, the closer AI technology approaches to a complete imitation of the human mind, the more relevant the current criminal law becomes.

Subjecting AI robots to the criminal law may relax our fears of human-robot coexistence. Criminal law plays an important role in giving people a sense of personal confidence. Each individual knows that all other individuals in society are bound to obey the law, especially the criminal law. If the law is breached by any individual, society enforces it by means of its relevant coercive powers (police, courts, and so on). If any individual or group is not subject to the criminal law, the personal confidence of the other individuals is severely harmed because those who are not subject to the criminal law have no incentive to obey the law.

The same logic works for humans, corporations, and AI entities.[74] If any of these is not subject to criminal law, the personal confidence of the other entities is harmed, and more broadly, the sense of confidence of the entire society is harmed. Consequently, society must make all efforts to subject all active entities to its criminal law. At times this requires conceptual changes in the general insights of society. For example, when the fear of corporations that are not subject to criminal law became real, in the seventeenth century, criminal law was applied to them. It is perhaps time to do the same for AI entities in order to alleviate human fears.[75]

But one more issue must be clarified. There are other creatures living among us, in our society, in coexistence with us: animals. Why not apply our legal rules concerning animals to AI entities as well?

1.2.3. Inapplicability of the Zoological Legal Model to AI Technology

Not all creatures on earth are considered by humans to be intelligent. Most of those considered to be intelligent are mammals. Nevertheless, humans and animals have coexisted for many millennia. Since humanity first domesticated animals for its own benefit, this coexistence has become intensive. We eat their products (milk, eggs, and so on); feed on their flesh; are protected by them (for example, guard dogs); employ them (in agriculture, transportation, and so on); and use them for both industrial and private needs. Therefore, the law has addressed the issues of our coexistence since ancient times.[76]

Examination of both ancient and modern law regarding animals reveals two important aspects that the law addresses. The first is the law's relation to animals as a property of humans, and the second is our duty to show mercy toward animals. The first aspect involves the ownership, possession, and other property rights of humans toward animals. For example, if damage is caused by an animal, the human who has the property rights over the animal is legally responsible for the damage. If a person is attacked by a dog, its owner is legally responsible for any damage.

In most countries these are issues of tort law, although in some countries they are related to criminal law as well, but in either case the legal responsibility is the human's, not the animal's. Since ancient times, an animal was incapacitated only if it was considered too dangerous for society. Incapacitation usually meant killing the animal. This was the case if an ox gored a person in ancient times,[77] and it still is the case if a dog bites a person under certain circumstances. No legal system considers an animal to be directly subject to the law, especially not to criminal law, regardless of the animal's intelligence.

The second aspect of law—the duty to show mercy toward animals—is aimed at humans. As humans are considered superior to animals because of their human intelligence, animals are viewed as helpless. Consequently, the law prohibits the abuse of power by humans against animals because of cruelty. The subjects of these legal provisions are humans, not animals. Animals abused by humans have no standing in court, even if damaged. The legal "victim" in these cases is society, not the animals, so in most cases, these legal provisions are part of criminal law.

Society indicts humans who abuse animals because such cruelty harms society, not the animals. This type of protection of animals differs little from the protection of property. Most criminal codes prohibit damaging another's property, in order to protect the property rights of the possessor or owner. But in the case of cruelty to animals, this type of protection has nothing to do with property rights. The legal owner of a dog may be indicted for abusing the dog regardless of property rights with respect to the dog. These legal provisions, which have existed since time immemorial, form the zoological legal model. The question is why this model cannot be applied to AI entities.

We are considering three types of entities: humans, animals, and artificially intelligent entities. If we wish to subordinate AI entities to the criminal law of humans, we must be able to justify the resemblance between humans and AI entities in this context. Indeed, we should explain why AI entities carry more resemblance to humans than to animals. Otherwise, the zoological legal model would be satisfactory and adequate for settling the AI activity. Thus, the question is, who does an AI robot resemble more: a human or a dog?

The zoological legal model has been examined previously with respect to AI entities when discussing the control of unmanned aircrafts[78] and other machines,[79] including "new generation robots."[80] For some of the legal issues raised in these contexts, the zoological legal model was able to provide answers, but the core problems could not be solved by this model.[81] When the AI entity was able to figure out by itself, using its software, how it needed to act, something in the legal responsibility puzzle was still missing. Communication of complicated ideas is much easier with AI entities than with animals. This is the case regarding external knowledge and the quality of reasonable conclusions in various situations.

An AI entity is programmed by humans to conform to formal human logical reasoning. This forms the core reasoning on which the activity of the AI entity is based, so that its calculations are accountable through formal human logical reasoning. Most animals, in most situations, lack this type of reasoning. It is not that animals are not reasonable, but their reasonableness is not necessarily based on formal human logic. Emotionality plays an important role in the activity of most living creatures, both animals and humans, and it may supply the drive and motivation to some human and animal activity. This is not the case with machines.

If emotionality is used as a measure, humans and animals are much closer to each other than either is to machines. But if pure rationality is used as a measure, machines may be closer to humans than to animals. Although emotionality and rationality affect each other, the law distinguishes between them regarding their applicability. For the law, especially for criminal law, the main factor being considered is rationality, whereas emotionality is only rarely taken into account. For example, a murderer's feelings of hatred, envy, or passion are immaterial for his or her conviction; and only in rare cases, when the provocation defense is used, may the court consider the emotional state of the offender.[82]

Further, the zoological legal model educates humans to be merciful toward animals, as already noted, which is a key consideration under this model. In relation to AI entities, however, this consideration is immaterial. Because AI entities lack basic attributes of emotionality, they cannot be saddened, made to suffer, disappointed, or tortured in any emotional manner, and this aspect of the zoological legal model has no significance regarding AI entities. For example, cutting off a dog's leg for no medical reason is considered abuse, and in some countries it is considered an offense. But no country considers cutting off a robot's arm an offense or abuse, even if the robot is a humanoid, and even if the robot is programmed to respond to this action "emotionally."[83]

Therefore, given that the law prefers rationality over emotionality when evaluating legal responsibility, and because the rationality of AI technology is based on formal human logical reasoning, AI technology is much closer to humans than to animals from the point of view of the law and with regard to all legal aspects. So the zoological legal model is not suitable for evaluating the legal responsibility of AI entities.[84] To subject AI entities to the criminal law, it is necessary to use the basic concepts of criminal liability, which form the general requirements of criminal liability and must be met in order to impose criminal liability on any individual (human, corporate, or AI) entity.

1.3. THE MODERN OFFENDER

Who is considered an offender under modern criminal law? Most people associate the terms *criminal* and *offender* with "evil." Murderers and rapists are criminals; therefore, they are evil. Within the confines of criminality, however, we encounter not only murder and rape, but other behaviors as well, which are not considered evil by most people. Some "white-collar" crimes are not always considered evil, and many people regard them as rather clever ploys or as succumbing to complex bureaucratic systems. Similarly, most breaches of traffic laws are not regarded as evil, nor are some of the offenses committed under the pressures imposed by discharging one's duty.

For example, consider the case of a surgeon who conducts a complicated emergency surgery to save a patient's life, and must work under great time pressure, so that the operation fails despite the surgeon's earnest efforts, and the patient dies. The post mortem examination reveals that one of the surgeon's acts may be considered negligent, and she becomes criminally liable for negligent homicide. Do we consider her evil?

Indeed, even some of the classic crimes, such as murder, do not always appear evil under certain circumstances. For example, a son witnesses the daily suffering of his mother, who is dying of a terminal illness; she asks him to disconnect her from the CPR machine to end her suffering. If the son reluctantly agrees, he is criminally liable for murder.[85] But is he evil? And what if he refuses his mother's request, in order to watch and presumably revel in her suffering? He would not be criminally liable for any offense, but would we not consider him evil?

Countless similar examples could be cited. Evil is not a measure of criminal liability. Morality of any kind and criminal liability are vastly different. At times they coincide, but such coincidence is not necessary for the imposition of criminal liability. An offender is a person on whom criminal liability has been imposed. When the legal requirements of a given criminal offense are met by individual behavior, criminal liability is imposed, and the individual is considered an offender. Evil may or may not be involved. Thus, imposition of criminal liability requires the exploration of its requirements.

Criminal law is considered the most efficient means of social control. Society, as an abstract body or entity, controls its individual members using a variety of measures (for example, moral, economic, or cultural), but one of the most efficient is the law. And given that only criminal law includes significant sanctions (imprisonment, fines, and even death), it is considered the most efficient of the means used to control individuals; this is commonly called *legal social control*. Imposition of criminal liability is the application and implementation of legal social control.

Modern criminal liability is independent of morality of any kind, and independent of evil. It is imposed in an organized, almost mathematical way. Two cumulative types of requirements must be met for the imposition of criminal liability: the first has to do with the law, the second with the offender. Both types of requirements must be met in order to impose criminal liability, and if they are met, no additional conditions are required.

1.3.1. Basic Requirements for a Given Offense

The first type of requirement, placed on the law, includes four separate requirements that every offense must meet, as defined by law. If the offense fails to meet even one of these requirements, no court can impose criminal liability on individuals for that offense. The four requirements are (1) legality, (2) conduct, (3) culpability, and (4) personal liability. Each of these four requirements represents a fundamental principle in criminal law, that is, the principle of legality, the principle of conduct, the principle of culpability, and the principle of personal liability.

Legality is a requirement for an offense to be considered legal (*nullum crimen sine lege*). It is legality that forms the rules used to determine what is criminally "right" and "wrong." For an offense to be considered legal, it must meet the following four cumulative conditions: (1) legitimate legal source, (2) applicability in time, (3) applicability in place, and (4) legitimate interpretation.[86] Only when all four conditions are met is the offense considered legal.

The offense must have a legitimate legal source that creates and defines it.[87] In most countries, the ultimate legitimate legal source is legislation, and case law is not considered a legitimate source because legislation is enacted by elected public representatives of society as a whole. Given that criminal law exercises legal social control, it should reflect the will of society by means of its representatives. The requirement that the offense be applicable in time means that retroactive offenses are illegal.[88] In planning their moves, individuals must know the prohibitions in advance, not retroactively. Only in rare cases are retroactive offenses considered legal, for example when a new offense is to the benefit of the defendant (it is more lenient than its predecessor or allows for a new defense) or when the offense covers cogent international custom, or *jus cogens*, (for example, genocide, crimes against humanity, or war crimes).[89]

The offense must also be applicable in place, so that extraterritorial offenses are illegal.[90] Criminal law is based on the authority of the sovereign, which is domestic, meaning that criminal law must be domestic as well. For example, the criminal law of France is applicable in France but not in the United States. Thus, extraterritorial offense is illegal (for example, French offense that is applicable in the United States). In rare cases, however, the sovereign is authorized to protect itself or its inhabitants abroad through extraterritorial offenses, as, for example, in the case of the foreign terrorists who attacked the US Embassy in Kenya and who may be indicted in the United States under US criminal law, although they have never been to the United States.[91] Extraterritorial offenses may also apply in cases of international cooperation between states.

Offenses must be well formulated and phrased. They must be general and address an unspecified public (for example, "Samuel Jackson is not allowed to . . ." is not a legitimate offense).[92] Offenses must be feasible, so that legal social control is realistic (for example, "Whoever breathes for more than five minutes uninterruptedly shall be guilty of . . ." is not a legitimate offense).[93] They must also be clear and precise, so that individuals will know exactly what they are allowed to do and what is prohibited conduct.[94] When all these conditions are met, the requirement of legality is satisfied.

Conduct is a requirement that every offense must meet in order to be considered legal (*nullum crimen sine actu*). Modern society prefers freedom of thought and has no interest in punishing mere thoughts (*cogitationis poenam nemo patitur*). Effective legal social control cannot be achieved through mind policing, which is not enforceable. The conduct is the objective and external expression of the commission of the offense.

Offenses that do not meet this are not legitimate. Throughout legal human history, only tyrants and totalitarian regimes have used offenses that lack conduct.

The conduct requirement of some offenses can be satisfied by inaction. Offenses that criminalize the status of the individual rather than his or her conduct are considered *status offenses*. For example, offenses that punish the relatives of traitors simply because they are relatives, regardless of their conduct, are considered status offenses.[95] So are offenses that punish individuals of certain ethnic origin.[96] Most developed countries have abolished these offenses, and defendants indicted for such offenses are acquitted in court because status offenses contradict the principle of conduct in criminal law.[97] Only when conduct is required may offenses be considered legal and legitimate.

The culpability requirement (*nullum crimen sine culpa*) must be met for an offense to be considered legal. Modern society has no interest in punishing accidental, thoughtless, or random events, only events that are the result of an individual's culpability. Someone's death does not necessarily indicate the presence of an offender. A person may pass next to another exactly when the other person is struck dead by lightning, but the passerby is not necessarily culpable. The offense must require some level of culpability for the imposition of criminal liability; otherwise, this would amount to cruel maltreatment of individuals by society.

Culpability relates to the mental state of the offender and reflects the subjective-internal expression of the commission of the offense. The required mental state of the offender, which forms the requirement of culpability, may be reflected both in the particular requirements of the specific offense and in the general defenses. For example, the offense of manslaughter requires recklessness as its minimal level of culpability.[98] But if the offender is insane (general defense of insanity),[99] is an infant (general defense of infancy),[100] or has acted in self-defense (general defense of self-defense),[101] no criminal liability is imposed. No culpability can be ascribed to such an offender. An offense is considered legal and legitimate only if it requires culpability.

Personal liability is required for an offense to be considered legal.[102] Modern society has no interest in punishing one person for the behavior of another, regardless of their relationship. Effective legal social control cannot be achieved unless all individuals are liable for their own behavior. If someone knows that no legal liability is imposed on him for his behavior, he has no incentive to avoid committing offenses or other types of antisocial behavior. Legal social control can be effective only when a person knows that for his own behavior, no other person is liable. Punishment deters individuals only if they may be punished personally.

Personal liability guarantees that each offender would be criminally liable and punished only for his or her own behavior. When individuals collaborate in committing an offense, each of the accomplices is criminally liable only for his or her own part. The accessory is criminally liable for accessoryship, and the joint perpetrator for joint perpetration. The various types of criminal liability in conjunction with the principle of personal liability have established the general forms of complicity in criminal law (for example, joint perpetration, perpetration-through-another, conspiracy, incitement, and accessoryship). Only when personal liability is required can the offense be considered legal and legitimate.

When all four basic requirements of legality, conduct, culpability, and personal liability are met, the offense that embodies them is considered to be legitimate and legal. Only then can society impose criminal liability on individuals for the commission of these offenses. The legitimacy and legality of specific offenses is necessary for the imposition of criminal liability, but not sufficient. For the imposition of criminal liability, the particular requirements of the offense must be met by the offender as well. These requirements are part of the definition of the offense.

1.3.2. Requirements of a Given Offense:
Modern Criminal Liability as a Matrix of Minimalism

The second type of requirement is focused on the offender's behavior. Each offense that meets the requirements to be considered such, determines the requirements needed for the imposition of criminal liability based on that offense. Although different offenses have differing requirements, the formal logic behind all offenses and their structure is similar. The common formal logic and structure are significant attributes of modern criminal liability. In general, these attributes are characterized by *minimalism*, which in this context means that the offense determines only the lower threshold for the imposition of criminal liability. In other words, the offender is required to meet *at least* the requirements of the given offense. There are two general requirements for an offense: (1) the factual element requirement and (2) the mental element requirement.

The modern structure of the factual element requirement is the same in most developed countries. This structure applies the fundamental principle of conduct to criminal liability, and it is identical for every type of offense, regardless of its mental element requirement. The factual element requirement is the broad objective-external basis of criminal liability (*nullum crimen sine actu*),[103] designed to answer four main questions about the factual aspects of the delinquent event: "What?" "Who?" "When?" and "Where?" "What" refers to the substantive facts of the event (what happened). "Who" relates to the identity of the offender. "When" addresses the time of the event. "Where" specifies the location of the event.

In some offenses, these questions are answered directly within the definition of the offense. In others, some of the questions are answered through the applicability of the principle of legality in criminal law.[104] For example, the offense "Whoever assaults another person . . ." does not relate directly to the questions of "who," "when," and "where," but the questions are answered through the applicability of the principle of legality. Because the offense is likely to be general,[105] the answer to the "who" question is *any person who is legally competent*. As this type of offense may not be applicable retroactively,[106] the answer to the "when" question is *from the time the offense was validated onward*. And because this type of offense may not be applicable extraterritorially (with some exceptions),[107] the answer to the "where" question is *within the territorial jurisdiction of the sovereign* (with some exceptions).

Nevertheless, the answer to the "what" question must be incorporated directly into the definition of the offense. This question addresses the core of the offense and cannot be answered through the principle of legality. This approach is at the foundation of the modern structure of the factual element requirement, which consists of three main components: conduct, circumstances, and results. Conduct is a mandatory component, whereas circumstances and results are not. Thus, if the offense is defined as having no conduct requirement, it is not legal, and the courts cannot convict individuals based on such a charge. Table 1.1 lists the components that answer the four questions.

The conduct component is at the heart of the answer to the "what" question. Status offenses, in which the conduct component is absent, are considered illegal, and in general they are abolished when discovered.[108] But the absence of circumstances or of results in the definition of an offense does not invalidate the offense.[109] These components are aimed at meeting the factual element requirement with greater accuracy than by conduct alone. Therefore, there are four formulas that can satisfy the factual element requirement:

1. conduct
2. conduct + circumstances
3. conduct + results
4. conduct + circumstances + results

Table 1.1 · Target Questions of the Factual Element Components

FACTUAL ELEMENT COMPONENTS	QUESTIONS
conduct	"what"
circumstances	"what," "who," "when," "where"
results	"what"

The modern structure of the mental element requirement has been widely accepted in most developed countries. This structure applies the fundamental principle of culpability in criminal law (*nullum crimen sine culpa*). The principle of culpability has two main aspects: positive and negative. The positive aspect (what needs to be in the offender's mind in order to impose criminal liability) relates to the mental element, whereas the negative aspect (what should *not* be in the offender's mind in order to impose criminal liability) relates to the general defenses.[110] For example, imposition of criminal liability for manslaughter requires recklessness as mental element, but it also requires that the offender *not* be insane. Recklessness is part of the positive aspect of culpability, and the general defense of insanity is part of its negative aspect.

The positive aspect of culpability in criminal law concerns the involvement of mental processes in the commission of the offense. The mental element has two important aspects: cognition and volition. Cognition is the individual's awareness of the factual reality. In some countries, awareness is called *knowledge*, but in this context there is no substantive difference between awareness and knowledge, both of which can relate to data from the present or the past, but not from the future.[111] A person may assess or predict what will be in the future, but not *know* or be *aware* of it. Prophetic skills are not required for criminal liability. Cognition in criminal law refers to a binary situation: the offender either is aware of fact X or is not. Partial awareness is not accepted in criminal law, and it is classified as unawareness.

Volition has to do with the individual's will, and it is not subject to factual reality. An individual may wish for unrealistic events to occur or to have occurred, in the past, the present, or the future. Volition is not binary because there are different levels of will, the three basic ones being positive (A wants X), neutral (A is indifferent toward X), and negative (A does not want X). The cognitive and volitive aspects combine to form the mental element requirement, as derived from the positive aspect of culpability in criminal law. In most developed countries, there are three principal forms of mental element, which are differentiated based on the cognitive aspect. The three forms represent three layers of positive culpability: (1) *mens rea* (general intent), (2) negligence, and (3) strict liability.

The highest layer of the mental element is that of *mens rea*, which requires full cognition. The offender is required to be fully aware of the factual reality. This layer involves examination of the offender's subjective mind. Negligence, the second layer, is cognitive omission: the offender is not required to be aware of the factual element, although based on objective characteristics, she could and should have had awareness of it. Strict liability is the lowest layer of the mental element; it replaces what was formerly known as absolute liability. Strict liability is a relative legal presumption of negligence based on the factual situation alone, which may be refuted by the offender (*praesumptio juris tantum*). Cognition relates the factual reality, as just noted. The relevant factual reality in criminal law is that which is reflected by the components of the factual element.

From the perpetrator's point of view, however, when committing the offense, only the conduct and circumstance components of the factual element exist in the present; the results occur in the future. Cognition is restricted to the past and the present, and therefore it relates to conduct and circumstances. Although results occur in the future, the possibility of their occurrence ensuing from the relevant conduct exists in the present, so that cognition can also relate to the possibility of the occurrence of the results. For example, in the case of homicide, A aims a gun at B and pulls the trigger. At this point, A is aware of his conduct, of the existing circumstances, and of the possibility of B's death as a result of his conduct.

Volition is considered immaterial for both negligence and strict liability, and can be added only to the mental element requirement of *mens rea*, which embraces all three basic levels of will. Table 1.2 compares the general requirements of the three layers, or forms, of mental element.

Because in most legal systems, the default requirement for the mental element is *mens rea*, negligence and strict liability offenses must specify explicitly the relevant requirement. The explicit requirement may be listed as part of the definition of the offense or included in the explicit legal tradition of interpretation. If no explicit requirement of this type is mentioned, the offense is classified as a *mens rea* offense, which is the default requirement. The relevant requirement may be met not only by the same form of mental element, but also by a higher-level form. Thus, the mental element requirement of the offense is the minimal level of mental element needed to impose criminal liability;[112] a lower level is insufficient for im-

Table 1.2 · Schematic Comparison of the Requirements of Mental Element Forms

	REQUIREMENT	
Form	Cognition	Volition
Mens rea	subjective cognition	only in result offenses (because of the relevant level)
Negligence	cognitive omission	none
Strict liability	presumed cognitive omission	none

Table 1.3 · Schematic Comparison of the Requirement for Satisfying the Mental Element Forms

MENTAL ELEMENT REQUIREMENT OF...	IS SATISFIED THROUGH...
mens rea negligence strict liability	*mens rea* negligence or *mens rea* strict liability or negligence or *mens rea*

posing criminal liability for the offense. Table 1.3 compares schematically the requirements for satisfying the three forms of mental element.

This structure of criminal liability forms a matrix of minimalist requirements. Each offense embodies the minimum requirement for the imposition of criminal liability, and meeting these requirements is sufficient to impose criminal liability. No additional psychological elements are required, so that the modern offender is an individual who meets the minimal requirement of an offense. The offender is not required to be "evil," "mean," or "wicked"—only to meet all the requirements of the offense. The imposition of criminal liability is therefore dry and rational (as opposed to emotional), and resembles mathematics.

Thus, the matrix of minimalist requirements for criminal liability has two aspects: structural and substantive. For example, if the mental element of the offense requires only awareness, then no other component of the mental element is required (structural aspect), and the required awareness is defined by criminal law irrespective of its meaning in psychology, philosophy, theology, and so on (substantive aspect). The substantive aspect is discussed in detail in chapters 2, 3, and 4.

Nevertheless, it is clear that the structure of criminal liability has been designed for humans, with the abilities of humans—and not other creatures—in mind. The mental element requirement relies on the human spirit, soul, and mind. Can machines, therefore, be examined based on human standards of spirit, soul, and mind? Considering these insights, how can criminal liability be imposed on entities lacking in spirit and soul?

It is important to remember, however, that although insights in criminal liability rely on the human spirit and soul, the imposition of criminal liability itself does not depend on these terms that carry deep psychological meaning. If an entity, any entity, meets both the factual and mental element requirements of the offense, then criminal liability may be imposed with or without spirit, with or without soul. This understanding is not new, and perhaps it is not highly innovative in the twenty-first century. A similar understanding was reached in the seventeenth century, long before the emergence of modern AI technology, but at a time when other nonhuman creatures committed offenses and it was necessary to subject them to criminal law. These creatures have neither spirit nor soul, but they have the ability to be subjects of criminal liability, which is being imposed on them to this day. These creatures are corporations.

1.3.3. The Case of Nonhuman Corporations (Round 1)

Since the seventeenth century, corporations have been among the modern offenders, although they lack spirit, soul, and a physical body.[113] Corporations were recognized already by Roman law, but the evolution of the modern corporation began in the fourteenth century, when English law demanded permission from the king or Parliament to recognize a corporation as legal.[114] Early corporations in the Middle Ages were mostly ecclesiastical bodies that were active in the organization of church property. From these bodies evolved associations, commercial guilds, and professional guilds that eventually formed the basis for commercial corporations.

During the sixteenth and seventeenth centuries, hospitals and universities were also commonly incorporated.[115] In addition to these, commercial corporations developed to solve problems of division of ownership among several owners of a business.[116] When people created a new business, ownership could be divided between them by establishing a corporation and dividing the shares and stocks among the shareholders. This pattern of ownership division has been perceived as efficient in minimizing the risks of the owners regarding the financial hazards of the business, and as a result, corporations have become common.[117]

The use of corporations flourished during the first industrial revolution, to the point where they have been identified both with the fruits of the revolution and with the misery of the lower classes and the workers (who were created by the revolution). Corporations were regarded as being responsible for the poverty of the workers, who did not share in the profits, and for the continuing abuse of children employed by the corporations. As the industrial revolution progressed, public and social pressure on the corporations increased, until legislators were forced to restrict the activities of corporations.

In the beginning of the eighteenth century, the British Parliament enacted statutes against the abuse of power by corporations, the very power granted to them by the state for the benefit of social welfare.[118] To ensure the effectiveness of these statutes, they included criminal offenses. The offense most often used for this purpose was that of public nuisance.[119] This legislative trend deepened as the revolution progressed, and in the nineteenth century, most developed countries had advanced legislation regarding the activities of corporations in a variety of contexts. To be effective, this legislation included criminal offenses as well, prompting the conceptual question, how can criminal liability be imposed on corporations?

Criminal liability requires a factual element, whereas corporations possess no physical body. Criminal liability also requires a mental element, whereas corporations have no mind, brain, spirit, or soul.[120] Some European countries refused to impose criminal liability on nonhuman creatures, and revived the Roman rule whereby corporations were not subject to criminal liability (*societas delinquere non potest*). But this approach was highly problematic because it created legal shelters for offenders. For example, an individual who did not pay taxes was criminally liable, but when the individual was a corporation, it was exempt. This provided an incentive for individuals to work through corporations and evade paying taxes. Eventually, all countries subjected corporations to criminal law, but not until the twentieth century.

The Anglo-American legal tradition accepted the idea of the criminal liability of corporations early because of its many social advantages and benefits. In 1635, criminal liability for corporations was enacted and imposed for the first time.[121] This was a relatively primitive imposition of criminal liability because it relied on vicarious liability, but it enabled the courts to impose criminal liability on corporations in a way that was separate from the criminal liability of any of its owners, workers, or shareholders. This structure remained in effect throughout the eighteenth and nineteenth centuries.[122]

The major disadvantage of criminal liability based on vicarious liability was that it required valid vicarious relations between the corporation and another entity, which in most cases happened to be human.[123] Consequently, when the human entity acted without permission (*ultra vires*), the corporation was exempt of responsibility. To exempt a corporation of liability, it was sufficient to include a general provision in its incorporation papers prohibiting the commission of any criminal offense on behalf of the corporation.[124] This deficiency resulted in the replacement in Anglo-American legal systems of the model of criminal liability for corporations during the late nineteenth century and early twentieth century.[125]

The new model was based on the *identity theory*. In some types of cases, the criminal liability of corporations derives from its organs, whereas in other types of cases, criminal liability is independent. When the criminal offense requires an *omission* (for example, not paying taxes, not fulfilling legal requirements, or not observing workers' rights), and the duty to act is the corporation's, the corporation is criminally liable independently, regardless of any criminal liability of other entities, whether human or not. When the criminal offense requires an *act*, the acts of its organs are related to the corporation if they were committed on its behalf, with or without permission.[126] The same structure works for the mental element, for *mens rea*, negligence, and strict liability alike.[127]

Consequently, the criminal liability of the corporation is direct, not vicarious or indirect.[128] If all requirements of an offense are met by the corporation, it is indicted regardless of any proceedings that may be in effect against any human entity. If convicted, the corporation is punished separately from any human entity. Punishments for corporations are considered to be not less effective than for humans. To complete this picture, we will return to discuss the case of nonhuman corporations (Round 2) in relation to the question of the sentencing of AI entities.[129]

However, the main significance of the modern legal structure of criminal liability of corporations is conceptual. Since the seventeenth century, criminal liability has not been a uniquely human domain. Other entities, nonhuman, are also subject to the criminal law. Some adjustments were needed to make this legal structure applicable, but eventually, nonhuman corporations have become subject to criminal law. Today, this seems entirely natural, as it should. Given that the first barrier was crossed in the seventeenth century, the road may be open to crossing the next barrier that stands before the imposition of criminal liability on AI entities.

In the following chapters, the criminal liability of AI entities is analyzed in detail for all types of offenses. Because the major obstacle on the way to criminal liability for AI entities is that of the mental element, the offenses are divided according to their mental element requirements (*mens rea*, negligence, and strict liability). Next, the applicability of general defenses for AI entities is analyzed. Finally, in the last chapter, the applicability of human punishments to AI entities is discussed.

2

AI CRIMINAL LIABILITY FOR INTENTIONAL OFFENSES

In 1981, a thirty-seven-year-old Japanese employee in a motorcycle factory was killed by an AI robot working near him. The robot identified the employee as a threat to its mission and calculated that the most efficient way to eliminate the threat was to push the worker into an adjacent machine. Using its very powerful hydraulic arm, the robot smashed the surprised worker into the operating machine, killing him instantly, after which it resumed its duties without further interference.[1] The legal question is, who is to be held criminally liable for this homicide?

Intentional offenses make up most of the criminal offenses in most countries. To impose criminal liability in these cases, both factual element and mental element requirements must be met by the apparent offender. Consequently, the question of the criminal liability of AI entities for intentional offenses is whether AI entities have the capability to meet these requirements accumulatively. Note, however, that even if AI entities have such capability, this does not exempt humans from their criminal liability, if humans are involved in the commission of the offense.

To explore these issues, we examine the capability of AI entities to meet the requirements of both factual and mental elements of intentional offenses to allow the imposition of human criminal liability for intentional AI offenses.

2.1. THE FACTUAL ELEMENT REQUIREMENT

The factual element requirement structure (*actus reus*) is identical for all types of offenses: intentional, negligence, and strict liability, as previously noted.[2] This structure contains one mandatory component (conduct) for all offenses, and two optional components (circumstances and results) for some offenses. Let us now discuss the capability of machines to meet the factual element requirement considering this structure.

2.1.1. Conduct

Conduct, as already noted, is the only mandatory component of the factual element requirement; in other words, an offense that does not require conduct is illegal and illegitimate, and no criminal liability can be legitimately imposed based on it. In independent offenses, the conduct component of the factual element requirement may be expressed both by an act and by omission. In derivative criminal liability, the conduct component may be expressed also by inaction, with some restrictions. We will examine these forms of conduct in relation to the capabilities of AI machines.

In criminal law, act is defined as material performance with a factual-external presentation. According to this definition, the materiality of the act is manifest in its factual-external presentation, which differentiates the act from subjective-internal matters related to the mental element. Because thoughts have no factual-external presentation, they are related not to the factual, but to the mental element. Will may initiate acts, but in itself has no factual-external presentation, and is therefore considered part of the mental element. Consequently, involuntary and unwilled actions, as well as reflexes, are still considered acts as far as the factual element requirement is concerned.[3]

But although unwilled acts or reflexes are considered acts, criminal liability is not necessarily imposed for such offenses because of reasons involving the mental element requirement or general defenses. For example, B physically pushes A in the direction of C. Although the physical contact between A and C is involuntary for both, it is still considered an act. But it is unlikely that criminal liability for assault will be imposed on A, because the mental element requirement for assault has not been met. Acts committed as a result of loss of self-control are still considered acts, but loss of self-control is a general defense, which if proven and accepted, exempts the offender from criminal liability.

This definition of an act concentrates on its factual aspects to the exclusion of mental aspects.[4] The definition is also sufficiently broad to include actions originating in telekinesis, psychokinesis, and so on, if these are possible,[5] as long as they have a factual-external presentation.[6] If act were restricted to "willed muscular construction" or "willed bodily movement,"[7] this would prevent imposition of criminal liability in cases of perpetration-through-another (for example, B in the example just cited) because no act has been performed, and it would be possible to assault someone and be exempt from criminal liability, by pushing an innocent person upon the victim.

Consequently, criminal law considers as an act any material performance with a factual-external presentation, whether willed or not. On this basis, AI technology is capable of performing acts that satisfy the conduct requirement. This is true not only for strong AI technology, but also for much lower technologies. When a robot moves its arms or any other of its devices, it is considered to act. This is true when the movement is the result of inner calculations carried out by the robot, but not only then. Even if the robot is fully actuated by a human operator through remote control, any movement of the robot is considered an act.

Therefore, even non–AI technology robots have the factual capability of performing acts, regardless of the motives or reasons for the act. This does not necessarily mean that these robots are criminally liable for their acts, because the imposition of criminal liability must also satisfy the mental element requirement. But as far as performing an act in order to satisfy the conduct component requirement is concerned, any material performance with a factual-external presentation is considered an act, whether the physical performer is a strong AI entity or not.

In criminal law, omission is defined as inaction contradicting a legitimate duty to act. According to this definition, the term *legitimate duty* is of great significance. The opposite of action is not omission, but inaction. If doing something is an act, then not doing anything is inaction. Omission is an intermediate degree of conduct between action and inaction. Omission is not mere inaction, but is inaction that contradicts a legitimate duty to act: the omitting offender is required to act, but fails to do so.[8] If no act has been committed, but no duty to act is imposed, no omission has been committed.[9]

The omitting offender is, therefore, punished for doing nothing in specific situations where he or she should have done something owing to a certain legitimate duty. For example, in most countries parents have a legal duty to care for their children. In these countries, the breach of this duty may form an offense. In this situation, the parent is not punished for *acting* in a wrong way, but for *not acting* although having a legal duty to act in a specific way. The requirement to act must be legitimate in the given legal system, and in most legal systems, the legitimate duty may be imposed both by law (legal duty) and by contract (contractual duty).[10]

The modern concept of conduct in criminal law acknowledges no substantive or functional differences between acts and omissions for the imposition of criminal liability.[11] Therefore, any offense may be committed both by act and by omission. Socially and legally, commission of offenses by omission is no less severe than commission by act.[12] Most legal systems accept this modern approach, and there is no need to explicitly require omission to be part of the factual element of the offense. The offense defines the prohibited conduct, which may be committed both through acts and through omissions.[13]

On this basis, AI technology can perform omissions that satisfy the conduct requirement. This is true not only for strong AI technology, but also for much lower technologies. Physically, commission by omission requires doing nothing. There is no doubt that any machine is capable of doing nothing; therefore, any machine is physically capable of committing an omission. Naturally, for the inaction to be considered omission, there must be a legal duty that contradicts the inaction. If such duty exists, originating in law or contract, and the duty is addressed to the machine, then there is no question that the machine is capable of committing an omission with respect to that duty.

This is also the situation regarding inaction. Inaction is the complete factual opposite of an act. If an act is doing something, then inaction is not doing it, or doing nothing. Whereas omission is inaction contradicting a legitimate duty to act, inaction requires no such contradiction. Omission is not doing when there is an obligation to do, whereas inaction is not doing when there is no obligation to do anything. Inaction is accepted as a legitimate form of conduct only in derivative criminal liability (for example, attempt, joint perpetration, perpetration-through-another, incitement, or accessoryship), and not in complete and independent offenses.[14]

In these instances, when inaction is accepted as a legitimate form of conduct, it is physically committed in the same way as omission. Consequently, if AI technology is capable of commission through omission of conduct, it is also capable of commission through inaction. This does not necessarily mean that machines or robots are automatically criminally liable for these omissions or inactions, because to impose criminal liability, it is necessary to also meet the mental element requirement, and it is not sufficient to satisfy the factual element requirement.

Thus, the mandatory component of the factual element requirement (conduct) can be satisfied by machines. These machines need not be highly sophisticated or even based on AI technology. Simple machines are capable of conduct under the definitions and requirements of criminal law. For the imposition of criminal liability on any entity, this is an essential step, even if not a sufficient one. No criminal liability can be imposed if the conduct requirement is not satisfied, but conduct alone is not sufficient for the imposition of criminal liability.

2.1.2. Circumstances

Circumstances are not a mandatory component of the factual element requirement. Some offenses require circumstances in addition to conduct, and some do not. Circumstances are defined as factual components that describe the conduct but do not derive from it. According to this definition, circumstances specify the criminal conduct in more accurate terms. When defining specific offenses, circumstances are required especially when the conduct component is too broad or vague, and greater specificity is needed in order to avoid over-criminalization of situations that are considered legal by society.

In the case of most offenses, the circumstances represent the factual data that make the conduct criminal. For example, in most legal systems, the conduct in the offense of rape is having sexual intercourse (although the specific verbs may vary). But in itself, having sexual intercourse is not an offense, and it becomes one only if it is committed without consent. The factual element of "without consent" is what makes the conduct of "having sexual intercourse" criminal. In this offense, the factual component "without consent" functions as a circumstance.[15] Additionally, the factual component "with a woman" also functions as circumstance because it describes the conduct (raping a chair is not an offense).

In this example, the factual element of the offense of rape is "having sexual intercourse with a woman without consent."[16] Whereas "having sexual intercourse" is the mandatory conduct component, "with a woman" and "without consent" function as circumstance components that describe the conduct more accurately, to make sure that the factual element requirement of rape is specified adequately, in a way that avoids over-criminalization.

According to the definition, circumstances are not derived from the conduct in order to allow distinguishing them from the result component.[17] For example, in homicide offenses, the conduct is required to cause the "death" of a "human being." Death describes the conduct and also derives from it because it is the conduct that caused death. Therefore, in homicide offenses, "death" does not function as a circumstance. The factual data that functions as a circumstance is "human being." The victim was a human being long before the conduct took place, and therefore it does not derive from the conduct. But these words also describe the conduct (causing the death of a human being, not of an insect), and therefore function as a circumstance.

On this basis, AI technology is capable of satisfying the circumstance component requirement of the factual element. When the circumstances are external to the offender, the identity of the offender is immaterial, and therefore the offender may be human or a machine, without affecting the circumstances. In the earlier example of rape, the circumstance "with a woman" is external to the offender. The victim is required to be a woman regardless of the identity of the rapist. The raped woman is still considered "woman" whether she was attacked by a human or a machine, or has not been attacked at all.

When the circumstances are not external to the offender but are related somehow to the offender's conduct, the circumstances assimilate within the conduct component. In the example of rape, the circumstance "without consent" describes the conduct as if it were part of the conduct (how exactly did the offender have sexual intercourse with the victim). To satisfy the requirement of this type of circumstance, the offender must commit the conduct in a particular way. Consequently, meeting the requirement of this type of circumstance is the same as committing the conduct. As already discussed, it is within the capabilities of machines to meet the conduct component requirement.

2.1.3. Results and Causal Connection

Results are not a mandatory component of the factual element requirement. Some offenses require results in addition to conduct, and some do not. Results are defined as a factual component that derives from the conduct. According to this definition, results specify the criminal conduct with greater accuracy. Results are defined as deriving from the conduct in order to allow distinguishing them from the circumstances. For example, in homicide offenses, the conduct is required to cause the "death" of a "human being." The death describes the conduct and also derives from it, because the conduct caused the death. Therefore, in homicide offenses, "death" does not function as a circumstance, but as a result.[18]

In the context of the factual element, the results derive from the conduct through factual causal connection. Although additional conditions exist for this derivation (legal causal connection), the factual requirement is a factual causal connection. Consequently, proof of results requires proving a factual causal connection.[19] The factual causal connection is defined as a derivative connection, in which, were it not for the conduct, the results would not have occurred the way they have. According to this definition, the results are the ultimate consequences of the conduct, that is, *causa sine qua non*.[20]

The factual causal connection relates not only to the mere occurrence of the results, but also to the way in which they occurred. For example, A hit B with her car, and B died. Because B was terminally ill, A may argue that B would have died anyway in the near future, so B's death is not the ultimate result of A's conduct, and the result would have occurred anyway, even without A's conduct. But because the factual causal connection has to do with the way in which the results occurred, the requirement is met in this example: B would not have died the way he did had A not run him over.

On this basis, AI technology is capable of satisfying the result component requirement of the factual element. To achieve the results, the offender must initiate the conduct. Commission of the conduct forms the results, and the results are examined objectively to determine whether they are derived from the commission of the conduct.[21] Thus, when the offender commits the conduct, and the conduct has been carried out, it is the conduct, not the offender, that is the cause of the results, if there was a result. The offender is not required to cause, separately, any results, but only to commit the conduct. Although the offender begins the factual process that forms the results, this process is initiated only through the commission of the conduct component.[22]

Because AI technology is capable of committing conducts of all types, in the context of criminal law it is capable of causing results deriving from this conduct. For example, when a robot operates a firearm and causes it to shoot a bullet toward a human individual, this fulfills the conduct component requirement of homicide offenses. At that point, a causal connection test is used to examine the conduct and determine whether it caused that individual's death. If it did, both the result and conduct component requirements are met, although physically the robot "did" nothing but commit the conduct. Because meeting the conduct component requirement is within the capabilities of machines, as already discussed, so is meeting the results component requirements.

2.2. THE MENTAL ELEMENT REQUIREMENT

The imposition of criminal liability for intentional offenses requires the fulfillment of both factual and mental element requirements. The mental element requirement of intentional offenses is general intent, more accurately known as *mens rea*, meaning "evil mind" in Latin. Because the term *general intent* can at times be confused with *intent* or *specific intent*, we prefer to use the more accurate term, *mens rea*. If AI technology is capable of fulfilling the *mens rea* requirement, it is possible, feasible, and achievable to impose criminal liability on AI entities for intentional offenses.

First, we explore the structure of the *mens rea* requirement, then we turn to the fulfillment of the requirement by AI technology.

2.2.1. Structure of the *Mens Rea* Requirement

In the modern age, *mens rea* expresses basic types of mental elements, as it embodies the idea of culpability most effectively. This is the only mental element that combines cognition and volition. The *mens rea* requirement expresses the internal-subjective relation of the offender with the factual commission of the offense.[23] In most legal systems, the *mens rea* requirement functions as the default option of the mental element requirement. Therefore, unless negligence or strict liability are explicitly required as mental elements of the offense, *mens rea* is the required mental element.

This default option is also known as the *presumption of mens rea*.[24] Accordingly, all offenses are presumed to require *mens rea*, unless there is some explicit deviation from this requirement. Because *mens rea* is the highest level of known mental element requirement, this presumption has great significance. Indeed, most offenses in criminal law require *mens rea* and not negligence or strict liability. None of the mental element components, including *mens rea* components, is independent or stands alone for itself. For example, the dominant component of *mens rea* is awareness. If we require the offender to be aware, the question arises, aware of what? Awareness cannot stand alone because it would be meaningless.

Consequently, all mental element components must relate to facts or to some factual reality. The relevant factual aspect for criminal liability consists, naturally, of the factual element components (conduct, circumstances, and results), as discussed earlier.[25] Factual reality contains many more facts than the components of factual element, but all other facts are irrelevant for the imposition of criminal liability. For instance, in the rape example, the relevant facts are "having sexual intercourse with a woman without consent,"[26] and the rapist must be aware of these facts. Whether the offender was or was not aware of other facts as well (for example, the color of the woman's eyes, the fact that she was pregnant, or her suffering) is immaterial for the imposition of criminal liability.

Thus, for the imposition of criminal liability, the object of the mental element requirement is the factual element components. Naturally, this object is much narrower than the complete factual reality, but the factual element represents the decision of society regarding what is relevant for criminal liability and what is not. Nevertheless, other facts and the mental relation to them may affect the punishment, but they are immaterial for the imposition of criminal liability. For example, a particularly cruel rapist would be convicted of rape, whether he was cruel or not, but his punishment would likely be much harsher than that of a less cruel rapist.

Identifying the factual element components as the object of *mens rea* components is the basis for the structure of *mens rea*. *Mens rea* has two layers of requirement: cognition and volition. The layer of cognition consists of awareness. The term *knowledge* is also used to describe the cognition layer, but awareness seems more accurate. Both awareness and knowledge function the same way and have the same meaning in this context. We can be aware only of facts that occurred in the past or are occurring at present, but we cannot be aware of future facts.

For example, we can be aware of the fact that A ate an ice-cream cone two minutes ago, and we can be aware of the fact that B is eating one right now. If C said that she intends to eat an ice-cream cone, we can predict it, foresee it, or estimate the probability that it will happen, but we cannot be *aware* of it, simply because it has not occurred yet. If the criminal law had required offenders to be aware of future facts, it would have required prophetic skills. As far as we know, most criminals do not possess such skills and are far from being prophets.

In this context, the offender's temporal point of view determines the point at which the conduct is actually performed. Therefore, from the offender's point of view, the conduct component always occurs in the present. Consequently, awareness is a relevant component of *mens rea* in relation to conduct. Circumstance is defined as factual data that describes the conduct but does not derive from it.[27] To describe current conduct, circumstances must exist in the present as well. For example, the circumstance "with a woman" in the offense of rape, described earlier, must exist simultaneously with the conduct "having sexual intercourse."

The raped person must be a woman during the commission of the offense for the circumstance requirement to be fulfilled. Therefore, awareness is a relevant component of *mens rea* in relation to circumstances as well. The situation is different, however, in relation to results, which are defined as a factual component that derives from the conduct; this means that results must occur after the conduct.[28] For example, B dies at 10:00:00 after which, at 10:00:10, A shoots her. In this case, the conduct (the shooting) is not the cause of the other factual event (B's death), which does not function as a result.

From the offender's point of view, given that the results occur after the conduct, and because the offender's temporal point of view determines the point at which the conduct is performed, the results occur in the future. Consequently, because results do not occur in the present, awareness is not relevant for them. The offender is not expected to be *aware* of the results, which have still not occurred from his point of view. But although the offender cannot be aware of the future results, he is capable of predicting them and assessing the probability of their occurrence. These capabilities are present simultaneously with the actual performance of the conduct.

For example, A shoots B. At the point in time when the shot is fired, B's death has not yet occurred, but in the act of shooting, A is aware of the possibility of B's death as a result of the shot. Therefore, awareness is a relevant component of *mens rea* in relation to the possibility that the results will occur, but not in relation to the results themselves. Awareness of this possibility is not required to be related to how reasonable or probable the occurrence of the result is. If the offender is aware of the mere existence of the possibility, whether its probability is high or low, that the result will follow from the conduct, this component of *mens rea* is present.[29]

The additional layer of *mens rea* is that of volition. This layer is additional to cognition and based on it. Volition is never alone, but always accompanies awareness. Volition relates to the offender's will regarding the results of the factual event. In rare offenses, volition may relate to motives and purposes beyond the specific factual event, and is expressed as specific intent.[30] The main question regarding volition (apart from the offender's awareness of the possibility that the results will follow from the conduct) is whether the offender *wanted* the results to occur. Because from the offender's point of view the results occur in the future, they are the only reasonable object of volition.

From the offender's temporal point of view, neither circumstances nor conduct have anything to do with will. The raped person was a woman before, during, and after the rape, regardless of the rapist's will. The sexual intercourse is what it is at that point in time, regardless of the rapist's will. If the offender argues that the conduct occurred against his will—that is, the offender did not control it—this argument is related to the general defense of loss of self-control, which we will discuss in a moment.[31] Thus, as far as conduct and circumstances are concerned, no volition component is required, only awareness.

In the factual reality there are many levels of will, but criminal law accepts only three: (1) intent (and specific intent), (2) indifference, and (3) rashness. The first represents positive will (the offender wants the results to occur), the second represents nullity (the offender is indifferent about the results), and the third represents negative will (the offender does not want the results to occur, but has taken unreasonable risks that caused them to occur). For example, in homicide offenses, at the moment when the conduct is committed, if the offender

1. wants the victim to die, it is intent (or specific intent);
2. is indifferent as to whether the victim lives or dies, it is indifference;
3. does not want the victim to die, but assumes unreasonable risk in this regard, it is rashness.

Intent is the highest level of will accepted by criminal law. In most countries, intended homicide is considered murder. Indifference is the intermediate level, and rashness is the lowest level of will. Both indifference and rashness are known as *recklessness*. In most countries, reckless homicide is considered manslaughter. Consequently, if the offense requires recklessness, this requirement can be fulfilled by proof of intent because a higher level of will satisfies lower levels. But if the offense requires intent or specific intent, this requirement can be fulfilled only by intent or specific intent.

It is easier to sum up the structure of *mens rea* if we divide offenses into *conduct offenses* and *result offenses*. Conduct offenses are those whose factual element requires no results; result offenses are those whose factual element requires results.[32] This division clarifies the structure of *mens rea* because volition is required only in relation to results. Therefore, results require both cognition and volition, whereas conduct and circumstances require only cognition. The structure of *mens rea* can be described schematically as shown in table 2.1.

Therefore, in conduct offenses, whose factual element requirement contains conduct and circumstances, the *mens rea* requirement contains awareness of these components. In result offenses, whose factual element requirement contains conduct, circumstances, and results, the *mens rea* requirement contains awareness of the conduct, of the circumstances, and of the possibility that the results will occur. In addition, in relation to results, the *mens rea* requirement contains intent or recklessness, according to the definition of the offense.

This general structure of *mens rea* is a template that contains terms derived from the terminology of the sciences of the mind (awareness, intent, recklessness, and so on). To examine whether AI entities are capable of fulfilling the *mens rea* requirement of given offenses, the definition of these terms must be explored. Next we will discuss separately the cognitive and volitive aspects of *mens rea*.

Table 2.1 · General Structure of *Mens Rea*

TYPE OF OFFENSE	*MENS REA* REQUIREMENT	
	Cognition	Volition
Conduct offense	awareness of conduct + awareness of circumstances	—
Result offense	awareness of conduct + awareness of circumstances + awareness of the possibility of the occurrence of results	intent or recklessness (indifference/ rashness)

2.2.2. Fulfillment of the Cognitive Aspect

The cognitive aspect of *mens rea* contains awareness, and therefore the relevant question is whether AI technology is capable of consolidating awareness. Because the term *awareness* may have different meanings in different scientific spheres (for example, psychology, theology, or law), to answer this question we must examine the legal meaning of this term; specifically, we must examine awareness based on its legal definition in criminal law. Even if the legal meaning may differ from other meanings of the term, given that our discussion relates to criminal liability, only the legal meaning is relevant.

In criminal law, awareness is defined as perception of factual data by the senses, and its understanding.[33] The roots of this definition lie in the psychological understandings of the late nineteenth century. Previously, awareness was equated with consciousness, that is, the physiological-bodily state the mind is in when the human is awake. By the end of the nineteenth century, psychologists argued that awareness involves a much more complicated situation of the human mind.[34] In the 1960s, the modern understanding of the human mind began to consolidate toward the concept of the sum total of internal and external stimulations that the individual is aware of at a given point in time.

Consequently, it became apparent that the human mind is not constant but dynamic and regularly changing, and could be described as a flow of feelings, thoughts, and emotions ("stream of consciousness").[35] It was also understood that the human mind is selective, meaning that humans are capable of focusing their minds on certain stimuli while ignoring others, so that the ignored stimuli do not enter the human mind at all. If the human mind were to include all internal and external stimuli and had to pay attention to each one of them, it would be incapable of functioning.

The function of the human and animal sensory systems is to absorb the stimuli (light, sound, heat, pressure, and so on) and transfer them to the brain for the processing of factual information. Processing is executed in the brain internally until a relevant general image of the factual data is created. This process is called perception, which is considered one of the basic skills of the human mind.

Many stimuli are active at any one time. To enable the creation of an organized image of factual data, the human brain must focus on some of the stimuli and ignore others. This is accomplished through the process of attention, which permits the brain to concentrate on some stimuli while ignoring others. In practice, the other stimuli are not entirely ignored, but they exist in the background of the perception process. The nervous system still retains adequate vigilance that allows it to absorb other stimuli when the process of attention is at work.

For example, Archie is watching television and is focused on it. Edith calls him because she wants to speak with him. When she first calls, he does not react. When she calls him for the second time, much louder, he reacts, asks her what she wants, and says that he must go to the bathroom. When Archie is focused on the television, many of the stimuli in his environment are ignored (for example, the sound of his heartbeats, the smells from the kitchen, and the pressure in his bladder), allowing him to focus on the television. But his nervous system is still sufficiently alert to absorb some of these stimuli. Edith's first call is just another sound to be ignored by the process of attention. But when she calls the second time, the attention process used to focus on the television is stopped, and the other stimuli are absorbed and receive some attention. This is why Archie suddenly remembers that he must go to bathroom.

Perception includes not only absorbing stimuli, but also processing them into a relevant general image, which is what usually gives meaning to the accumulation of stimuli. Processing the factual data into a relevant general image is accomplished through unconscious inference, so that there is absolutely no need for awareness to carry this out.[36] But the result of the processing—that is, the relevant general image—is a conscious result. Thus, although the human mind is not conscious of most of the process, it is conscious of its results when the relevant general image is accepted. Consequently, the human mind is considered to be aware only when the relevant general image is accepted.

This process forms the essence of human awareness, which is the final stage of perception. Perception of factual data by the senses, and its understanding, results in the creation of a relevant general image, which is the awareness of the factual data. For example, it is not the eyes but the brain that is the sight organ of humans. Eyes function as mere sensors that deliver factual data to the brain. Only when the brain creates the relevant general image is the human considered to be aware of the relevant sight. Consequently, a human in a vegetative situation, with fully functioning eyes, is not considered to be seeing (or to be aware of what she sees), unless the sights are combined into a relevant general image.

As a result, for us to consider a human to be aware of certain factual data, two accumulative conditions are required: (1) absorbing the factual data by senses, and (2) creating a relevant general image about these data in the brain. If one of these conditions is missing, the person is not considered to be aware. Awareness is a binary issue: the person either is aware or is not. Partial awareness is meaningless. An offender may be aware of part of the factual data, that is, fully aware of some of the data, but not partly aware of a certain fact. If the facts were not absorbed, or if no relevant general image has been created, the offender is considered to be unaware.

At times, the term *knowledge* is used to describe the cognitive aspect of *mens rea*, as already noted. Is there a difference between knowledge and awareness? When examined functionally, there is no difference between these terms as far as criminal liability is concerned because they both refer to the same idea of cognition.[37] Moreover, knowledge has been explicitly defined as awareness at times.[38] But it seems that within the context of the mental element, the more accurate term is awareness rather than knowledge.

Outside of the criminal law context, awareness is more related to consciousness than is knowledge. Knowledge is also related to the cognitive process, but it refers more to information than consciousness does. Knowledge represents perhaps a deeper cognitive process than awareness, but within the specific context of the mental element requirement in criminal law, awareness seems to be the more accurate term.

In court, proving the full awareness of the offender beyond any reasonable doubt, as required in criminal law, is not an easy task. Awareness relates to internal processes of the mind, which do not necessarily have external expressions. Therefore, criminal law has developed evidential substitutes for this task. These substitutes are presumptions, which in certain types of situations presume the existence of awareness. Two major presumptions are recognized in most legal systems: (1) the willful blindness presumption, as a substitute for awareness of conduct and of circumstances; and (2) the awareness presumption, as a substitute for awareness of the possibility of the occurrence of the results. Before exploring these presumptions and their relevance to AI technology, we must examine how AI technology meets the cognitive aspect of the *mens rea* requirement.

Is an AI robot capable of being aware of conduct, circumstances, or the possibility of the occurrence of the results in the context of criminal law?[39] The process of awareness may be divided into two stages, as noted. The first stage consists of absorbing the factual data by means of the senses. At this stage, the primary role is played by the devices used to absorb the factual data. The human devices are organs, such as the eyes (for sensing light), the ears (for sensing sound), and so on. These organs absorb the factual data (such as sights, light, sounds, pressure, and texture) and transfer it to the brain for processing.

AI technology has the capability and the relevant devices to absorb any factual data that can be collected by the five human senses. Cameras absorb light and transfer the data to processors,[40] microphones pick up sounds,[41] transducers sense pressures, thermometers sense temperature, hygrometers sense humidity, and so on. Indeed, advanced technologies are capable of much more accurate sensing than their corresponding human organs. For example, cameras may absorb light waves at frequencies that the human eye cannot detect, and microphones can sense sound waves that are inaudible to human ears.

How many of us can guess the exact outside temperature and humidity within 0.1 degrees? What about 0.0001 degrees? And what about relative humidity? Most of us cannot. By contrast, even the simplest electronic sensors can perform accurate measurements of these data and transfer the information to the appropriate processors. AI technology, therefore, is capable of meeting the requirements of the first stage of awareness, indeed much better than humans.

The second stage of the awareness process consists of creating a relevant general image about these data in the brain, that is, full perception. Naturally, most AI technologies, robots, and computers do not possess biological brains, but they possess artificial ones. Most of these "brains" are embodied in the hardware (processors, memory, and so on) of the technology they use. Are these brains capable of creating a relevant general image out of the factual data they absorb? Humans create the relevant general image by analyzing the factual data, which enables them to use the information, transfer it, integrate it with other information, and act based on it—in other words, to understand it.[42]

Consider the example of security robots based on AI technology.[43] The task of these robots is to identify intruders and either alert human troops (police, army) or stop the intruders on their own. Their sensors (cameras, microphones) absorb the factual information and transfer it to the processors, which are then supposed to identify the intruder. To do so, the processors analyze the factual data, making sure not to mistake police and other security personnel in the area for intruders. The analysis is based on identifying and discriminating between changing sights and sounds. The robots may compare the shape and color of clothes and use whatever other attributes they need to make the identification.

Following this short process, the robot assesses the probabilities. If the probabilities do not produce an accurate identification, the robot initiates a process of vocal identification and asks the suspicious figure to identify itself, provide a password, or perform any other action relevant to the situation. If the intruder answers, the sound is compared with sound patterns stored in memory. After all the necessary factual data has been collected, the robot has sufficient information to make a decision, on the basis of which it can act. Indeed, this robot has created a relevant general image out of the factual data absorbed by its sensors, which it was able to use, transfer, and integrate with other information, and then act based on it—in other words, to understand it.

Consider how a human guard would act in this situation: very likely, in the same way. The human guard would identify a figure or a sound as suspicious, whereas the robot would examine a change in the current image or sound. We prefer robot guards because they work more thoroughly than their human counterparts and do not fall asleep on the job. Both the human guard and the robot use their memories to try to identify the figure or the sound by comparing it with patterns stored in memory. But whereas the robot cannot forget figures, humans can. The human guard is not always sure, but the robot assesses all the probabilities. The human guard calls out to ask the figure to identify itself, and so does the robot. Both the human and the robot compare the answer with information stored in memory—yet another function that the robot carries out more accurately than its human counterpart. Following all the data collection, both human and robot guards make a decision. They both understand the situation.

We say that the human guard was aware of the relevant factual data. Can we not say the same about the robot guard? There is no reason why not. Its internal processes are more or less the same as those of the human guard, except that the robot is more accurate, faster, and works more thoroughly than the human guard. When they become aware of a suspicious figure or sound, they both absorb the data and act accordingly.

It is possible to argue that the human guard may have absorbed much more factual information than the robot—for example, smells—and that a human is capable of filtering out irrelevant information, such as cricket sounds. This argument does not invalidate the analysis just presented. Robots are also capable of absorbing additional factual data, such as smells. But as already noted, humans use the attention process to focus on a *portion* of the factual data, as Archie did in the example cited earlier. Although humans are capable of absorbing a wide range of factual data, this may prove to be a distraction for attending to the task at hand. Robots may be programmed to ignore such data.

If smells are considered irrelevant for a task of guarding, then robots will not consider this data. If they were human, we would say that the robots will not *pay attention* to this data. Humans filter out irrelevant data using the process of attention that runs in the background of the human mind. Robots, however, examine all factual data and eliminate the irrelevant options only after analyzing the data thoroughly. We may now ask ourselves which type of guard we humans would prefer: one who unconsciously does not pay attention to factual data, or one who examines all factual data thoroughly?

In sum, AI technology is also capable of fulfilling the requirements for the second stage of awareness. And given that the two stages already described are the only stages of the awareness process recognized by criminal law, we may conclude that AI technology is capable of meeting the awareness requirement in criminal law.

Nevertheless, some of us may have the feeling that something is still missing before we conclude that machines are capable of awareness. This may be correct if we consider awareness in its broader sense, as it is used in psychology, philosophy, or the cognitive sciences. But in this book we are examining the criminal liability of AI entities, and not the wide meaning of cognition in psychology, philosophy, the cognitive sciences, and so on. Therefore, the only standards of awareness that are relevant here are the standards of criminal law. The other standards are irrelevant for the assessment of criminal liability imposed on either humans or AI entities.

The criminal law definition of awareness is indeed much narrower than corresponding definitions in the other spheres of knowledge. This is true for the imposition of criminal liability not only on AI entities, but also on humans. The definitions of awareness in psychology, philosophy, or the cognitive sciences may be relevant for the endless quest for *machina sapiens*, but not for *machina sapiens criminalis*, which is based on the definitions of criminal law.[44] This distinction is one of the best illustrations of the difference between *machina sapiens* and *machina sapiens criminalis*.

As noted earlier, awareness itself is very difficult to prove in court, especially in criminal cases where it must be proved beyond any reasonable doubt.[45] Therefore, criminal law developed two evidential substitutes for awareness: the presumptions of willful blindness and awareness. According to the willful blindness presumption, the offender is presumed to be aware of the conduct and circumstances if he or she suspected that they existed but did not verify that suspicion.[46] The rationale of this presumption is that because the "blindness" of the offender to the facts is willful, he or she is considered to be aware of these facts even if not actually aware of them.

If the offender wished to avoid the commission of the offense, he or she would have checked the facts. For example, the rapist who suspects that the woman does not consent to having sexual intercourse with him predicts that if he were to ask her, she would refuse, in which case there would be no doubt about her lack of consent. Therefore, he ignores his suspicion, and at his interrogation says that he thought she consented because he heard no objection. The willful blindness presumption equates unchecked suspicion with full awareness of the relevant conduct or circumstances. The question is whether this presumption is relevant to AI technology.

According to the awareness presumption, a person is presumed to be aware of the possibility of the occurrence of the natural results of his or her conduct.[47] The rationale of this presumption is that humans have the basic skills necessary to asses the natural consequences of their conduct. For example, when firing a bullet at someone's head, the shooter is presumed to be able to asses the possibility of death as a natural consequence of the shooting. Humans who lack such skills, permanently or temporarily, have an opportunity to refute the presumption. The question is whether this presumption is relevant to AI technology.

Awareness is indeed difficult to prove for humans. But because within a criminal law context, the processes that comprise awareness can be monitored accurately by AI technology, there is no need for any substitute. In human terms, this would be the same as if the brain of every individual were constantly being scanned and every thought recorded. If awareness of factual data could be identified on such a brain scan, proving awareness beyond any reasonable doubt would be a simple task.

Given that every act of an AI entity can be monitored and recorded, including all processes that comprise awareness within a context of criminal law, it is entirely achievable to prove the awareness of an AI entity regarding particular items of factual data. Therefore, it is not necessary to prove the awareness of an AI entity by means of substitutes, because awareness can be proven directly. But if it were necessary to prove awareness using substitutes, this could also be accomplished. Strong AI technologies use algorithms to assess probabilities in order to make reasonable decisions. A minimum probability defines the feasibility, probability, or "reasonableness" of every event. This praxis can be applied beneficially to awareness substitutes.

For humans, only a realistic suspicion is considered adequate for willful blindness. Thus, if the probability of a particular factual event occurring is above a certain rate and the AI entity ignores the possibility, this may be considered willful blindness. The reason a human would not check such a suspicion is immaterial for the applicability of the willful blindness presumption, and the same is true for AI entities. So if the argument is that the AI entity ignores the possibility out of some evil or concealed desire, as humans may do, the reason for this ignorance is immaterial, just as it is in the case of humans.

As noted earlier, evil is not one of the components of criminal liability. Thus, the reason for ignoring a suspicion (that is, an alternative with a probability higher than a certain threshold) is immaterial, regardless of whether the motive is evil. The natural consequence of certain conduct has a reasonable probability of occurring. A machine can assess reasonability quantitatively by weighing the relevant circumstances. Strong AI technology is capable of identifying the options that humans would call "natural." Thus, although proving the awareness of an AI entity is not subject to the difficulties of proving human awareness, and therefore there is no need to use the awareness substitutes, these substitutes can be used to prove the awareness of AI entities.

2.2.3. Fulfillment of the Volitive Aspect

The volitive aspect of *mens rea* contains three levels of will: intent, indifference, and rashness. As already noted, indifference and rashness are commonly referred to as recklessness. The question is whether AI technology is capable of consolidating these levels of will. Because these terms have different meanings in various scientific spheres (psychology, theology, philosophy, law, and so on), to answer this question we must examine the legal meaning of these terms; specifically, their legal definitions in criminal law. Even if the legal meaning of these terms differs from other meanings, given that the question relates to criminal liability, only the legal meaning is relevant.

Intent is the highest level of will accepted by criminal law. There is some confusion about the meanings of the terms *intent*, *general intent*, and *specific intent*.[48] General intent is a common term for *mens rea*. To avoid confusion with intent, we use the term *mens rea* and not general intent. The term *intent* refers to the highest level of will embodied in the volitive aspect of *mens rea*. The term *specific intent* refers also to the highest level of will. Intent and specific intent have identical levels of will. But intent refers to will toward the result components of the factual element, whereas specific intent refers to motives and purposes, not to results.[49] Specific intent is relatively rare.

Purpose is factual data that are supposed to derive from the conduct. The mission of the conduct (or its destiny) is to achieve the purpose. Purpose offenses do not require that the purpose actually be achieved, only that the offender intends to achieve the purpose. For example, "Whoever says anything for the purpose of intimidating . . ." is an offense that does not require actual intimidation of anyone, only the presence of such purpose. Motive consists of internal-subjective feeling, and its satisfaction is the conduct derived from it. For example, in "Whoever performs X out of hatred . . ." the conduct X satisfies the hatred.

Unless purpose and motive are explicitly required by the offense, they are immaterial for the imposition of criminal liability.[50] For example, (1) A killed B out of hatred, and (2) A killed B out of mercy. These are both considered murders because the offense of murder does not require specific intent to achieve a set purpose or any motives. But if the offense had explicitly required specific intent toward a set purpose or any motives, proving the specific intent would have been a condition for the imposition of criminal liability in that offense. Again, the requirement of specific intent is relatively rare.

Given that the only difference between intent and specific intent involves their objects (results versus motives and purposes), and that the level of will of both is identical, the following analysis of intent applies to specific intent as well. Intent is defined as "aware will" that accompanies the commission of conduct, the results of which will occur. In the definition of specific intent, results are replaced by motives and purposes as follows: aware will that accompanies the commission of conduct, the motive for which is satisfied or the purpose of which is achieved.

Intent is an expression of positive will, that is, the will to make a factual event occur.[51] Although there are higher levels of will than intent (for example, lust, longing, or desire), in criminal law intent has been accepted as the highest level of will required for the imposition of criminal liability.[52] Consequently, no offense requires a higher level of will than intent, and if intent is proven, it satisfies all other levels of will, including recklessness (indifference and rashness).

We must distinguish between aware will and unaware will. Unaware will is an internal urge, impulse, or instinct of which the person is not aware and which is naturally uncontrollable. Individuals cannot control their will unless they are aware of it. Being aware of one's will does not guarantee the ability of controlling it; but controlling the will requires awareness of it and the activation of conscious processes in the mind to cause the relevant activity to start, cease, or continue without interference. Imposing criminal liability based on intent requires an individual who is aware of his will so that he may control it.

Intent does not require some abstract aware will. The aware will must be focused on certain targets: results, motives, or purposes. Intent is an aware will focused on these targets. For example, the murderer's intent is an aware will focused on causing the victim's death. For the intent to be relevant to the imposition of criminal liability, this will must exist simultaneously with the commission of the conduct and must accompany it. If a person kills another by mistake, and then, after the victim's death, develops the will to kill the victim, this is not intent. Only when the murder is committed simultaneously with the existence of the relevant will can the will be considered intent.

Proving intent is much more difficult than proving awareness. Although both are internal processes of the human mind, intent relates to future factual situations, and awareness relates to current facts. Awareness is rational and realistic, but intent is not necessarily so. For example, a person may have the intention of becoming a fish, but she cannot have an awareness of *being* a fish. Because of the serious difficulties in proving intent, criminal law developed evidentiary substitutes for it. The most common substitute is the foreseeability rule (*dolus indirectus*), a legal presumption designed to prove the existence of intent.

According to the foreseeability rule presumption, the offender is presumed to intend the occurrence of the results if during commission of the conduct, with full awareness, he has foreseen the occurrence of the results as highly probable.[53] If the object of the results is replaced with purpose, this presumption applies also to specific intent.[54] The rationale for this presumption is as follows: believing that the probability of a specific factual event following from the conduct is extremely high, committing the conduct expresses the offender's desire that the factual event should occur.

For example, A holds a loaded gun pointed at B's head. A knows that if she pulls the trigger, there is a high probability that the factual event of B's death will occur. A pulls the trigger. A then argues in court that she did not want B's death to occur, and therefore the required component of intent has not been fulfilled. If the court exercises the foreseeability rule presumption, A is presumed to have intended the occurrence of the result because in A's assessment, the result of death was highly probable and she acted accordingly, presumably because she wanted these results to occur.[55] This presumption is commonly used in courts to prove intent. Indeed, unless the defendant confesses explicitly during the interrogation to having had the intent, the prosecution prefers to prove the intent using this presumption.

Is an AI robot capable of having intent in the context of criminal law?[56] Because *will* is a vague and general term, even in criminal law, an AI entity's capability of having intent must be examined through the lens of the foreseeability rule presumption. Indeed, this is the core reason for using the presumption in proving human intent. Two conditions must be met by this rule: (1) occurrence of the results must be foreseeable with a very high probability, and (2) the conduct must have been committed with full awareness.

As already noted, strong AI entities are capable of assessing probabilities of the occurrence of factual events and acting accordingly. For example, chess-playing computers can analyze the current status of the game based of the location of the pieces on board. These computers review all possible options for the next move, and for each option they review all the possible responses of the other player. Then, for each response they again review all possible responses, and so on until the final possible move that ends the game with a victory for one of the players. Each possible move is assessed for its probability, based on which, the computer decides on its next move.[57]

If the player were human, we would say that he has an intention of winning the game. We would not *know* with certainty that he has such an intention, but the course of his conduct is consistent with the foreseeability rule presumption. An AI computer programmed to play chess has a goal-driven behavior of winning chess games, as do human chess players. If we can say that human chess players have the intention of winning chess games, it seems that we can say the same thing about AI chess. Analysis of their course of conduct in the relevant situations is entirely consistent with the foreseeability rule presumption.

Any entity, human or AI, that examines several options for conduct and then makes an aware decision to commit one of them, while assessing the probability that a given factual event would result from this conduct with a high degree of probability, is considered to be foreseeing the occurrence of that factual event. In the case of a chess game, this seems quite normal to us. But there is no substantial difference, in this context, between playing chess for the purpose of winning and committing any other conduct for the purpose of bringing about some result. If the result and the conduct together form a criminal offense, we enter the territory of criminal law.

When a computer assesses the probability that a factual event (winning a chess game, for example, or killing or injuring a person) will result from its conduct and chooses accordingly to commit the relevant conduct (moving chess pieces on the board, pulling a trigger, or moving its hydraulic arm in the direction of a human body), the computer meets the required conditions for the foreseeability rule presumption. The computer is therefore presumed to have the intention that the results should occur. This is exactly how the court examines the offender's intent in most cases, when the offender does not confess.

One may ask, what is considered to be a very high probability in this context? The answer is the same as for humans, the only difference being that the computer can assess the probability more accurately than a human can. For example, one points a loaded gun at someone's head. A human shooter evaluates that the probability of the victim's death is high as a result of pulling the trigger, but most humans are incapable of assessing *exactly* how high the probability is. If the entity is a computer, it can assesses the *exact* probability of occurrence based on the factual data to which it is exposed to (the distance of the victim from the gun, the direction and velocity of the wind, the mechanical condition of the gun, and so on).

If probability is assessed to be high, the computer is required to act accordingly, and in full awareness to commit a conduct that advances the relevant factual event. As already noted, AI technology has the capability to consolidate awareness to factual data.[58] Commission of the conduct is considered as factual data, therefore AI technology can meet both conditions needed to prove the foreseeability rule presumption. This presumption is considered to be an absolute legal presumption (*praesumptio juris et de jure*); therefore, if its conditions are met, there is no way of refuting its conclusion (that is, having intent).

This analysis of the foreseeability of AI technology relies on strong AI, the more specific requirement being the advanced capability of assessing probabilities for the purpose of decision making. Because strong AI technologies have this capability, they are therefore capable of intent within the context of criminal law. Thus, AI technology can meet the intent requirement in criminal law.

Nevertheless, some of us may still have the feeling that something is missing before we conclude that machines are capable of intent. This may be correct if we consider intent in its broader sense, as used in psychology, philosophy, or the cognitive sciences. But here we are examining the criminal liability of AI entities, and not the wider meaning of cognition. Therefore, the only standards of intent that are relevant to our examination are the standards of criminal law. Other standards are irrelevant for the assessment of criminal liability imposed on either humans or AI entities.

The criminal law definition of intent (and foreseeability) is indeed much narrower than corresponding definitions in other spheres of knowledge. This is true for the imposition of criminal liability not only on AI entities, but also on humans. The definitions of intent in psychology, philosophy, or the cognitive sciences may be relevant for the endless quest for *machina sapiens*, but not for *machina sapiens criminalis*, which is based on the definitions of criminal law.[59] This distinction is one of the best illustrations of the difference between *machina sapiens* and *machina sapiens criminalis*.

The evidence concerning the intent of the AI entity is based on the ability to monitor and record all the activities of the computer. The activities of the computer with regard to each stage in the consolidation of intent or foreseeability are monitored and recorded. Assessing probabilities and making relevant decisions based on the assessment are part of the computer's activities, so there is always direct evidence for proving the criminal intent of an AI entity if the proof is based on the foreseeability rule presumption.

If intent is proven, either directly or through the foreseeability rule presumption, then all other forms of volition may be proven accordingly. Because recklessness, consisting of indifference or rashness, represents a lower degree of will, it may be proven either by proving recklessness directly or by proving intent. There are no offenses that require only recklessness and nothing else. Because criminal liability requirements function as a matrix of minimal requirements, as noted earlier,[60] every requirement represents the minimum condition for the imposition of criminal liability. The prosecution may choose between proving the given requirement or any higher (but not lower) one.

Consequently, offenses that require recklessness as their mental element requirement can be satisfied by proving either intent (directly or through the foreseeability rule presumption) or recklessness. Thus, if the given AI technology has the foreseeability capabilities already described, it can meet the mental element requirements of both intent and recklessness offenses. But AI technology can also meet the recklessness requirement directly. Analogously, if the capability for intent is present, the capability for recklessness, which is a lower one, is also present.

Indifference is a higher level than recklessness, and it represents a fully aware state of volitional neutrality toward the occurrence of the factual event. For the indifferent person, the significance of the factual event occurring and not occurring is exactly the same. This has nothing to do with the actual probability of the occurrence of the factual event, only with the offender's internal volition toward such occurrence. For example, A and B are playing Russian roulette.[61] When it is B's turn, A is indifferent to the possibility of B's death; A does not care whether B lives or dies.

For indifference to be considered as such in the context of criminal law, it must be aware; that is, the offender must be aware of the various options and have no preference for any of them.[62] The decision-making process in strong AI technology is based on assessing probabilities, as we have discussed. When the computer makes a decision to act in a certain way, but this decision does not take into consideration the probability that a given factual event can occur, the computer is indifferent to the occurrence of that factual event.

In general, complicated decision-making processes in AI technology are characterized by many factors to be considered. In similar situations, humans tend to ignore a portion of the factors and not consider them. Computers do the same. Some are programmed to ignore certain factors, but strong AI technology is capable of learning to ignore factors—otherwise, the decision-making process would be impossible. The learning process used by strong AI technology, machine learning, is inductive learning from examples. The more examples analyzed, the more effective the learning.[63] Humans call this "experience."

Whenever the decision maker is an AI entity, the decision-making process is monitored, so there is no evidential problem in proving the indifference of the AI entity toward the occurrence of a given factual event. The awareness of the AI entity to the possibility of the occurrence of the event is monitored, together with the factors that were taken into consideration in the decision-making process. This data provides direct proof of the indifference of the AI entity. Indifference can also be proven through the foreseeability rule presumption, as noted earlier.

Rashness is the lower level of recklessness. It consists of aware volition for the relevant factual event not to occur, but at the same time the commitment of aware and unreasonable conduct that causes it to occur. Although the occurrence of the factual event is undesired by the rash person, she conducts herself with unreasonable risk so that the event eventually occurs. Rashness is considered a degree of will, because if the rash offender had not wanted the event to occur at any price, she would not have taken unreasonable risks in her conduct. This is the major reason that rashness is part of the volitive aspect of *mens rea*. Otherwise, negative will alone does not justify criminal liability.

For example, someone drives on a narrow road behind a slow truck. It is a two-way road divided by a solid white line. The driver is in a hurry and eventually decides to pass the truck, crossing the solid line. She does not intend to kill anyone; all she wants is to pass the truck. But a motorcyclist happens to come from the opposite direction. The driver is unable to avoid the motorcyclist, and kills him. If the driver had wanted the motorcyclist to be dead, she would have been criminally liable for murder, but this was not the case. Nor would it be true to say that she was indifferent to the death of the motorcyclist. She was rash, however. She did not want to hit the motorcycle, but she took unreasonable risks that resulted in death. She is therefore criminally liable for manslaughter.

In the context of criminal law, rashness requires awareness. The offender is required to be aware of the possible options, to prefer not to cause the occurrence of the resulting event, but to commit conduct that is unreasonable if she were to avoid the occurrence of the event.[64] The decision-making process of strong AI technology is based on assessing probabilities, as we have discussed. When the computer makes a decision to act in a certain way, but it does not weigh one of the relevant factors as sufficiently significant, it is considered to be rash regarding the occurrence of a given factual event. In our example, if the car driver were replaced by a driving computer, and if the computer calculated the probability of colliding with the motorcycle by crossing the solid line to be low, the decision to pass would be considered rash.

As already noted, complicated decision-making processes in AI are characterized by many factors to be considered. In situations like this, humans sometimes miscalculate the weight of some factors. So do computers. Humans reinforce their decisions by hopes and beliefs, but computers do not use such methods. Some computers are programmed to weigh certain factors in a certain way, but strong AI technology can learn to weigh factors correctly and accurately. Again, this learning by strong AI is called machine learning.

Whenever the decision maker is an AI entity, the decision-making process is monitored; therefore, there is no shortage of evidence in proving rashness on the part of an AI entity regarding the occurrence of a given factual event. The AI entity's awareness of the possibility of the occurrence of the event is monitored, together with the factors that were taken into consideration in the decision-making process, and also the actual weight they were assigned in the given decision. This data provides direct proof of the rashness of the AI entity. As noted earlier, rashness can also be proven through the foreseeability rule presumption.

Therefore, all components of the volitive aspect of *mens rea* apply to AI technology and can be proven in court. So the question is, who is to be held criminally liable for the commission of this type of offense?

2.3. CRIMINALLY LIABLE ENTITIES FOR INTENTIONAL AI OFFENSES

In general, imposition of criminal liability for *mens rea* offenses ("intentional offenses") requires the fulfillment of both factual and mental elements of these offenses. Humans are involved in the creation of AI technology and entities, in their design, programming, and operation. Consequently, when an AI entity meets the factual and mental elements of the offense, who is to be held criminally liable? The possible answers are the AI entity, the humans involved with the AI entity, or both. Let us examine each of these answers.

2.3.1. AI Entity Liability

If a human offender meets both the factual and mental element requirements of an offense, criminal liability for that offense is imposed. In doing so, the court does not need to investigate whether the offender was "evil," or whether some other attribute characterized the commission of the offense. The fulfillment of these requirements is the only condition for imposing criminal liability. Other information may affect the punishment, but not the criminal liability. We prefer our criminal legal system to act in this way because *evil* and *good* are matters of perspective.[65] Different people with differing perspectives regard these terms in different ways.[66]

The factual and mental elements are neutral in this context, and do not necessarily have evil or good content.[67] Meeting their requirements is much more "technical" than the detection of evil would be. For example, we prohibit murder. Murder is causing the death of a human with the awareness and intent of causing death. If an individual factually causes another person's death, the factual element requirement is fulfilled. If the conduct was committed with full awareness and intent, the mental element is also fulfilled, and the individual is criminally liable for murder, unless one of general defenses is applicable (for example, self-defense or insanity). The reason for the murder is immaterial for the imposition of criminal liability, and it makes no difference whether the murder was committed out of mercy (euthanasia) or out of villainy.

This is the way that criminal liability is imposed on human offenders. If we embrace this standard in relation to AI entities, we can impose criminal liability on them as well. This is the basic idea behind the criminal liability of AI entities, and it differs from their moral accountability, social responsibility, or even civil legal personhood.[68] The narrow definition of criminal liability makes it possible for the AI entity to become subject to criminal law. Nevertheless, the reader may feel that something is still missing in this analysis, that perhaps it falls short. We try to refute these feelings using rational arguments.

One such feeling may be that the capacity of an AI entity to follow a program is not sufficient to enable the system to make moral judgments and exercise discretion, although the program may contain a highly elaborate and complex system of rules.[69] This feeling may relate eventually to the moral choice of the offender. The deeper argument is that no formal system can adequately make the moral choices that may confront an offender. Two answers apply to this argument.

First, it is not certain that formal systems are morally blind. There are many types of morality and moral values.[70] Teleological morality, such as utilitarianism, deals with the utility values of conduct. These values may be measured, compared, and decided upon based on quantitative comparisons. For example, an AI robot controls a heavy wagon that malfunctions, leaving two possible paths for the wagon to follow. The robot calculates the probabilities, and determines that following one path will result in the death of one person, and following the other path will result in the death of fifty people. Teleological morality would direct any human to choose the first path, and the person who followed that path would be considered a moral person. The robot can be directed to act in accordance with this morality. The robot's morality is dictated by its program, which makes it evaluate the consequences of each possible path. Do humans not act the same way?

Second, even if the first answer is not convincing, and formal systems are considered incapable of morality of any kind, criminal liability is still neither dependent on any morality nor fed by it. Morality is not even a condition for the imposition of criminal liability. Criminal courts do not asses the morality of human offenders when imposing criminal liability. The offender may be highly moral in the court's view, and still be convicted, as in the case of euthanasia, for example. Alternatively, the offender may be very immoral in the eyes of the court, and still be acquitted, as in the case of adultery. Given that no morality of any type is required for the imposition of criminal liability on human offenders, it should not be a consideration when AI entities are involved, either.

Some may feel that AI entities are not humans, and that criminal liability is intended for humans only because it involves constitutional human rights that only humans may have.[71] In this context, it is immaterial whether the constitutional rights refer to substantial rights or to procedural rights. The answer to this argument is that perhaps criminal law was originally designed for humans, but since the seventeenth century, it has not been applied to humans exclusively, as we have discussed.[72] Corporations, which are nonhuman entities, are also subject to criminal law, and not only to criminal law. For the past four centuries, criminal liability and punishments have been imposed on corporations, albeit with some occasional adjustments.

Others may argue that although corporations have been recognized as subject to criminal law, the personhood of AI entities should not be recognized because humans have no interest in recognizing it.[73] This argument cannot be considered applicable in an analytical legal discussion on the criminal liability of AI entities. There are many cases in everyday life in which the imposition of criminal liability does not benefit human society, and still criminal liability is imposed. A famous example comes from Immanuel Kant, who is interpreted as believing that if the last person on Earth is an offender, he should be punished even if it leads to the extinction of the human race.[74] Human benefit has not been recognized as a valid component of criminal liability.

Finally, some may feel that the concept of awareness presented here is too shallow for AI entities to be called into account, blamed, or faulted for the factual harm they may cause.[75] This feeling is based on confusion: the concept of awareness and consciousness in psychology, philosophy, theology, and cognitive sciences is confused with the concept of awareness in criminal law. In other spheres of knowledge, we lack a clear notion of what awareness is, making it impossible to offer coherent answers to the question of AI capability of awareness. Whenever such answers are given, they are usually based on intuition rather than on science.[76]

Criminal law, however, must be accurate. Based on definitions provided by criminal law, people may be jailed for life, and they may lose their property or even their lives. Therefore, criminal law definitions must be proven beyond any reasonable doubt. Imposition of criminal liability for *mens rea* offenses is based on awareness as the major and dominant component of the mental element requirement. The fact that the term *awareness* in psychology, philosophy, theology, and cognitive sciences has not been developed adequately in order to create an accurate definition does not exempt criminal law from developing such a definition of its own for the purpose of imposing criminal liability.

The criminal law definition of awareness, like all other legal definitions, may differ significantly from its common meaning or from its meaning in psychology, philosophy, theology, and cognitive sciences. Criminal law definitions are designed and adapted to meet the needs of criminal law and nothing more. These definitions represent the requirements necessary for the imposition of criminal liability. They also represent the minimal conditions, both structural and substantive, based on the concept of criminal law as a matrix of minimal requirements, as discussed earlier.[77]

Consequently, the definitions of criminal law, including the definition of awareness, are relative and relevant only to criminal law. Because they are formed in conformity with the matrix of minimal requirements, they may be regarded as shallow. But this is true only when examined from the perspectives of psychology, philosophy, theology, and cognitive sciences, not from that of criminal law. If a significant development were to occur in those scientific spheres with regard to awareness, the criminal law may embrace newer and more comprehensive definitions of awareness inspired by these scientific discoveries. For the time being, there are none.

Criminal law definitions, including that of awareness, were originally designed for humans. For centuries people were indicted, convicted, and acquitted based on these definitions. Since the seventeenth century, these definitions have been adopted and adapted to the incrimination of nonhuman entities as well, namely corporations, which have also been indicted, convicted, and acquitted based on these definitions. Whenever the criminal law definitions were changed, they were changed for both humans and corporations in the same way. In the twenty-first century, the same definitions are required for the incrimination of AI entities.

Examination of these definitions reveals that they are applicable to AI technology, which can meet the requirements of criminal law without a single change to these definitions. Does that suddenly make them "shallow," when the same definitions have been used by criminal law systems worldwide for centuries and were considered adequate and almost unassailable? If they are good enough for humans and corporations, why would not they be good enough for AI technology?

If criticism claimed that criminal law definitions are too shallow and must therefore be radically changed for humans, corporations, and AI technology, the argument might be acceptable. But when these definitions are called shallow only in relation to AI technology, the argument cannot be considered serious or applicable. Such criticism originates in the endless quest for *machina sapiens*, and it is part of that quest. But because criminal law is a limited field, the quest for *machina sapiens criminalis* ended successfully when AI entities showed that they are capable of complying with the definitions of criminal law.

This answer applies not only to the criticism regarding the awareness of AI entities in criminal law, but also to the intentionality of AI entities.[78] We can now summarize that the criticism of the idea of AI entities' criminal liability generally relates to two points: (1) an absence of attributes that are not required for the imposition of criminal liability (for example, soul, evil, and good); and (2) the shallowness of criminal law definitions from the perspective of science other than criminal law (for example, psychology, philosophy, theology, and cognitive sciences). Both points of criticism can be answered methodically.

The first criticism can be refuted by pointing out the structural aspect of the matrix of criminal law, and noting that none of the absent attributes is required for the imposition of criminal liability on humans, corporations, and AI entities alike. The second criticism is answered through the substantive aspect of the matrix of criminal law, where definitions of legal terms from outside the sphere of criminal law are entirely irrelevant for settling the question of criminal liability. This opens the door for the imposition of criminal liability on AI entities as direct offenders.

Acceptance of the idea of the criminal liability of AI entities for *mens rea* offenses does not end in the commission of a certain offense by AI entities as principal offenders. The commission of *mens rea* offenses can also take place through complicity. No less than *mens rea* is required for the imposition of criminal liability on accomplices, because there is no complicity through negligence or strict liability.[79] A joint perpetrator may be considered as such only in relation to *mens rea* offenses. Other general forms of complicity (for example, incitement and accessoryship) also require at least *mens rea*.

In sum, because all general forms of complicity require at least *mens rea*, an AI system may be considered as an accomplice only if it actually formed *mens rea*. Making it possible to impose criminal liability on AI entities as direct offenders also recognizes the possibility of AI entities being accomplices, joint perpetrators, inciters, accessories, and so on, as long as both factual and mental element requirements are met in full. But these are not the only possible criminal liabilities that can be imposed on AI entities. Two additional ways are possible: perpetration-through-another and probable consequence liability.

2.3.2. Human Liability: Perpetration-through-Another

Humans, corporations, and AI technology alike may be used as instruments in the commission of offenses, regardless of their legal personhood. For example, one person threatens another that he will be killed if he does not commit a certain offense. Having no choice, the threatened person commits the offense. Who is considered criminally liable for the commission of that offense? The threatening person? The threatened person? Both? In the context of the criminal liability of AI entities, this question arises when the AI entity is used as an instrument by another entity. In the future, perhaps, the other way around may be relevant as well . . .

For situations of this type, criminal law has created the general form of criminal liability called perpetration-through-another. This may be defined as aware execution of a criminal plan through instrumental use of another person, who participates in the commission of the offense as an *innocent agent* or *semi-innocent agent*. Perpetration-through-another is a late development of vicarious liability in a law of complicity. Vicarious liability has been recognized in both criminal and civil law since ancient times, and it is based on the ancient concept of slavery.[80]

The master, who was a legal entity and possessed legal personhood, was liable not only for his own conduct but also for that of all his subjects (slaves, workers, family, and so on). When one of these subjects committed an offense, it was considered as if the master himself had committed the offense, and the master was obligated to respond to the indictment (*respondeat superior*). The legal meaning of this obligation was that the master was criminally liable for offenses physically committed by his subjects. The rationale for this concept was that the master should enforce the criminal law among his subjects. If the master failed to do so, he was personally liable for the offenses committed by his subjects.

Because the master's subjects were considered his property, he was liable for the harms committed by them, under both criminal and civil law. A subject was considered an organ of the master—his long arm. The legal maxim that governed vicarious liability stated that whoever acts-through-another is considered to be acting for himself (*qui facit per alium facit per se*). The physical appearance of the commission of the offense was immaterial for the imposition of criminal liability in this context.

This legal concept was accepted in most ancient legal systems. Based on it, Roman law developed the function of the father of the family (*paterfamilias*), who was responsible for any crime or tort committed by members of the family, its servants, guards, and slaves.[81] Consequently, the father of the family was also responsible for the prevention of criminal offenses and civil torts among his subjects. His incentive for doing so was his fear of criminal or tort liability for actions committed by members of his household. The legal concept of vicarious liability was absorbed into medieval European law.

The concept of vicarious liability was formally and explicitly accepted in English common law in the fourteenth century,[82] based on legislation enacted in the thirteenth century.[83] Between the fourteenth and seventeenth centuries, English common law amended the concept and ruled that the master was liable for the servants' offenses (under criminal law) and torts (under civil law) only if he explicitly ordered the servants to commit the offenses, explicitly empowered them to do so, or consented to their doing so before the commission of the offense (*ex ante*), or after the commission of the tort (*ex post*).[84]

After the end of the seventeenth century, this firm requirement was replaced by a much weaker one. Criminal and civil liability could be imposed on the master for offenses and torts committed by the servants even if the orders of the master were implicit or the empowerment of the servants was general.[85] This was the result of an attempt by English common law to deal with the many tort cases against workers at the dawn of the first industrial revolution in England and of the commercial developments of that time. The actions committed by the master's workers were considered to be the actions of the master because he enjoyed their benefits. And if the master enjoyed the benefits of these actions, he should be legally liable for the harm that might be caused by them, under both criminal and civil law.

In the nineteenth century, the requirements were further weakened, and it was ruled that if the worker's actions were committed through or as part of the general course of business, the master was liable for them even if no explicit or implicit orders had been given. Consequently, the defense argument of the worker having exceeded his authority (*ultra vires*) was rejected. Thus, even if the worker acted in contradiction to the specific order of a superior, the superior was still liable for the worker's actions if they were carried out in the general course of business. This approach was developed in tort law, but the English courts did not restrict it to tort law, and applied it to criminal law as well.[86]

Thus, vicarious liability was developed under specific social conditions, under which only individuals of the upper classes had the required competence to be considered legal entities. In Roman law, only the father of the family could become a prosecutor, plaintiff, or defendant. When the concept of social classes began to fade, in the nineteenth century, vicarious liability faded away with it.

In criminal law at the beginning of the nineteenth century, the cases of vicarious liability fell into three main types. The first was classic complicity. If the relations between the parties were based on real cooperation, they were classified as joint perpetration even if they had an employer-employee or some other hierarchical relation. But if within the hierarchical relations, information gaps between the parties or the use of power made one party lose its ability to commit an aware and willed offense, the act was no longer considered joint perpetration. The party that had lost the ability to commit an aware and willed offense was considered an innocent agent, functioning as a mere instrument in the hands of the other party.

The innocent agent was not criminally liable, and the offense was considered perpetration-through-another, where another party had full criminal liability for the actions of the innocent agent.[87] This was the basis for the emergence of perpetration-through-another from vicarious liability, and it became the second type of criminal liability derived from vicarious liability. The third type accounted for the core of the original vicarious liability. In most modern legal systems, this type is embodied in specific offenses and not in the general formation of criminal liability. Since the emergence of the modern law of complicity, the original vicarious liability is no longer considered a legitimate form of criminal liability.

Since the end of the nineteenth century and the beginning of the twentieth, the concept of the innocent agent has been widened to include parties with no hierarchical relations between them. Whenever a party acts without awareness of its actions or without will, it is considered an innocent agent. The acts of an innocent agent could be the results of another party's initiative (for example, using the innocent agent through threats, coercion, misleading, or lies); or the results of another party's abuse of an existing factual situation that eliminates the awareness or will of the innocent agent (for example, abuse of a factual mistake, insanity, intoxication, or infancy).

During the twentieth century, the concept of perpetration-through-another has been applied also to semi-innocent agents, typically a negligent party that is not fully aware of the factual situation although any other reasonable person could have been aware of it under the same circumstances. Most modern legal systems accept the semi-innocent agent as part of perpetration-through-another, so that the other party is criminally liable for the commission of the offense, and the semi-innocent agent is criminally liable for negligence.

If the legal system contains an appropriate offense of negligence (that is, the same factual element requirement, but a mental element of negligence instead of awareness, knowledge, or intent), the semi-innocent agent is criminally liable for that offense. If no such offense exists, no criminal liability is imposed, although the other party is criminally liable for the original offense. For the criminal liability of perpetration-through-another, the factual element may be fulfilled by the innocent agent, but the mental element requirement must be fulfilled actually and subjectively by the perpetrator-through-another, including with regard to the instrumental use of the innocent agent.[88]

So the question arises, if an AI entity is used by another (human, corporation, or another AI entity) as an instrument for committing an offense, how would criminal liability be divided between them? Perpetration-through-another does not consider the AI entity that physically committed the offense as possessing any human attributes. The AI entity is considered an innocent agent, but we cannot ignore an AI entity's capabilities to physically commit the offense. These capabilities are not sufficient to consider the AI entity as the perpetrator of the offense, because the entity lacks the required awareness or will.

The capabilities of the AI entity to physically commit the offense resemble the corresponding capabilities of a mentally limited person, such as an infant,[89] a person who is mentally incompetent,[90] or one who lacks a criminal state of mind.[91] Legally, when an offense is committed by an innocent agent (an infant,[92] a person who is mentally incompetent,[93] or one who lacks a criminal state of mind to commit an offense[94]), no criminal liability is imposed on the physical perpetrator. In such cases, that person is regarded as an instrument, albeit a sophisticated one, while the party orchestrating the offense (the perpetrator-through-another) is the actual perpetrator as a principal in the first degree, and is held accountable for the conduct of the innocent agent.

The perpetrator's liability is determined on the basis of the conduct of the "instrument"[95] and of his mental state.[96] The derivative question regarding AI entities is, who is the perpetrator-through-another? The answer is any person who makes an instrumental use of the AI technology for the commission of the offense. In most cases, this person is human, and may be the programmer of the AI software, its operator, or its end user. A programmer may design the AI entity in order to commit offenses through it. For example, a programmer designs software for a robot for use in a factory, and its software is programmed to torch the factory at night when no one is present. The robot committed the arson, but the programmer is considered the perpetrator.

The end user did not program the software, but uses the AI entity, including its software, for her own benefit, which is expressed by the commission of the offense. For example, a user purchases a household robot designed to execute any order given by its master. The robot identifies the user as the master, and the master orders the robot to assault any intruder to the house. The robot executes the order exactly as instructed. This situation is not different from that of a person who orders her dog to attack any trespasser. The robot commits the assault, but the user is considered the perpetrator.

In both scenarios, the actual offense was physically committed by an AI entity. Because the programmer and the user did not perform any action that conforms to the definition of a specific offense, they do not meet the factual element requirement of the offense. The liability in perpetration-through-another considers the physical actions committed by the AI entity as if they had been committed by the programmer, the user, or any other person instrumentally using the AI technology. The legal basis for this criminal liability is the instrumental use of the AI entity as an innocent agent.[97] No mental attribute required for the imposition of criminal liability is credited to the AI entity.[98]

When programmers or operators use an AI entity instrumentally, the commission of an offense by the AI entity is attributed to them. The mental element required for the offense already exists in their minds. The programmer had criminal intent when he ordered the commission of the arson, and the user had criminal intent when she ordered the commission of the assault, even if these offenses were physically committed by a robot, which is an AI entity. When an end user makes instrumental use of an innocent agent to commit an offense, the end user is considered to be the actual perpetrator of the offense.

Perpetration-through-another does not attribute any mental or human mental capabilities to the AI entity. Therefore, there is no legal difference between an AI entity, a screwdriver, and an animal that are instrumentally used by the perpetrator. When a burglar uses a screwdriver to pry open a window, he does so instrumentally, and the screwdriver is not criminally liable. The "action" of the screwdriver is in practice the burglar's. This is the same legal situation as when using an animal instrumentally. An assault committed by a dog by order of its master is considered as an assault committed by the master.

This type of criminal liability may be suitable for two kinds of scenarios. In the first scenario, an AI entity, even strong AI technology, is used to commit an offense without taking advantage of the advanced capabilities of the entity. In the second scenario, a weak version of AI technology is used, which lacks the advanced capabilities of modern AI entities. In both scenarios the AI entity is used instrumentally, but nevertheless, in both cases the AI entity is used because of its ability to execute an order to commit the offense. A screwdriver cannot execute such an order, but a dog can. A dog cannot execute a complicated order, but an AI entity can.[99]

Liability of perpetration-through-another is not applicable when an AI entity makes the decision to commit an offense based on its own accumulated experience or knowledge, or based on advanced calculations of probabilities. This liability is not suitable when the software of the AI entity was not designed to commit a specific offense, which was committed by the AI entity nevertheless. This type of liability does not apply either when the AI entity functions not as an innocent, but as a semi-innocent agent.[100] Semi-innocent agents lack the *mens rea* component but have a lower mental element component, such as negligence or strict liability, as we will discuss later.[101]

Liability of perpetration-through-another may still be applicable, however, when a programmer or user makes instrumental use of an AI entity, but without using its advanced capabilities. The legal result of applying this liability is that the programmer and the user are criminally liable for the offense committed, whereas the AI entity has no criminal liability of any type.[102] This is not significantly different from treating AI personhood as mere property, albeit with sophisticated skills and capabilities.[103] If the AI entity is considered a semi-innocent agent, that is, it fulfills the negligence or strict liability requirements, it is criminally liable for offenses of negligence or strict liability if there are recognized by the criminal law.

2.3.3. Joint Human and AI Entity Liability: Probable Consequence Liability

The first type of criminal liability, presented earlier, treated the AI entity as the perpetrator of the offense.[104] The second treated the AI entity as a mere instrument in the hands of the legally-considered perpetrator.[105] This second type of liability is not the only one possible to describe the legal relations between humans and AI entities in the commission of the offense. The second type dealt with adhered AI entity. But what if the AI entity, which was not programmed to commit the offense, calculated its actions, which then turn out to be considered an offense? The question here concerns the human liability rather than that of the AI entity.

For example, the programmer of a sophisticated AI entity designs it *not* to commit certain offenses. In the beginning, after being placed in operation, the AI system commits no offenses. In time, machine learning through induction widens the repertoire of the AI entity, which pursues new paths of activity. At some point, an offense is committed. In another, somewhat different example, the programmer designs the AI system to commit one given offense. As expected, the AI system commits the offense, but then it deviates from the original plan of the programmer and continues its delinquent activities. The deviation may be quantitative (more offenses of the same type), qualitative (more offenses of a different type), or both.

If the programmer had designed the entity from the beginning to commit the additional offenses, this would have been considered perpetration-through-another at most. But this is not what the programmer did. If the AI system consolidates both factual and mental elements of the additional offenses, it is criminally liable according to the first type of liability. But the question here concerns the criminal liability of the programmer. This is the main issue of the third type of liability, which we will discuss.

The most appropriate criminal liability in these cases is the probable consequence liability. Originally, probable consequence liability in criminal law was related to the criminal liability of parties to a criminal offense that has been committed in practice, but that was not part of the original criminal plan. For example, A and B plan to commit a bank robbery. According to the plan, A's role is to break into the safe, and B's role is to threaten the guard with a loaded gun. During the robbery the guard resists, and B shoots him to death. The killing of the guard was not part of the original criminal plan. When the guard was shot, A was not there, did not know about it, did not agree to it, and did not commit the shooting.

The legal question in this example concerns A's criminal liability for homicide, in addition to his obvious criminal liability for robbery. A does not satisfy either the factual or the mental element requirements of homicide, because he did not physically commit it, nor was he aware of it. The homicide was not part of the perpetrators' joint criminal plan. The question may be expanded also to the inciters and accessories to robbery, if any. In general, the question of the probable consequence liability refers to the criminal liability of a person for unplanned offenses that were committed by another. Before applying the probable consequence liability to human-AI offenses, we must explore its features.[106]

There are two extreme opposite approaches to this general question. The first calls for imposition of full criminal liability on all parties; the second calls for broad exemption from criminal liability for any party that does not meet the factual and mental element requirements of the unplanned offense. The first is considered problematic for over-criminalization, while the second is considered problematic for under-criminalization. Consequently, moderate approaches were developed and embraced by various legal systems.

The first extreme approach does not consider at all the factual and mental elements of the unplanned offense. This approach originates in Roman civil law, which has been adapted to criminal cases by several legal systems. According to this approach, any involvement in the delinquent event is considered to include criminal liability for any further delinquent event derived from it (*versanti in re illicita imputantur omnia quae sequuntur ex delicto*).[107] This extreme approach requires neither factual nor mental elements for the unplanned offense from the other parties, besides the party that actually committed the offense and accounts for both factual and mental elements.

According to this extreme approach, the criminal liability for the unplanned offense is an automatic derivative. The basic rationale of this approach is deterrence of potential offenders from participating in future criminal enterprises by widening the criminal liability to include not only the planned offenses, but the unplanned ones as well. The potential party must realize that his personal criminal liability may not be restricted to specific types of offenses, and that he may be criminally liable for all expected and unexpected developments that are derived directly or indirectly from his conduct. Potential parties are expected to be deterred and avoid involvement in delinquent acts.

This approach does not distinguish between various forms of involvement in the delinquent event. The criminal liability for the unplanned offense is imposed regardless of the role of the offender in the commission of the planned offense as perpetrator, inciter, or accessory. The criminal liability imposed for the unplanned offense does not depend on meeting factual and mental element requirements by the parties. If the criminal liability for the unplanned offense is imposed on all parties of the original enterprise, including those who could have no control over the commission of the unplanned offense, the deterrent value of this approach is extreme.

Prospectively, this approach educates individuals to keep away from involvement in delinquent events, regardless of the specific role they may potentially play in the commission of the offense. Any deviation from the criminal plan, even if not under the direct control of the party, is basis for criminal liability for all individuals involved, as if it had been fully perpetrated by all parties. The effect of this extreme approach can be broad and encompassing. Parties to another (third) offense, different from the unplanned offense, who were not direct parties to the unplanned offense, may be criminally liable for the unplanned offense as well, if there is the slightest connection between the offenses. The criminal liability for the unplanned offense is uniform for all parties and requires no factual and mental elements. Most Western legal systems consider such a deterrent approach too extreme, and have rejected it.[108]

The second extreme approach is the exact opposite of the first, and focuses on the factual and mental elements of the unplanned offense. According to this approach, to impose criminal liability for the unplanned offense, both factual and mental element requirements must be met by each party. Only if both requirements of the unplanned offense are met by a given party is it legitimate to impose criminal liability on it. Naturally, because the unplanned offense was not planned, it is most unlikely that any of the parties would be criminally liable for that offense, except for the party that actually committed it. This extreme approach ignores the social endangerment inherent in the criminal enterprise, which includes not only planned offenses, but unplanned ones as well.

In light of this extreme approach, offenders have no incentive to restrict their involvement in the delinquent event. Prospectively, any party that wishes to escape criminal liability for the probable consequences of the criminal plan needs only to avoid participation in the factual aspect of any further offense. Such offenders would tend to involve more parties in the commission of the offense in order to increase the chance for committing further offenses, and therefore most modern legal systems prefer not to adopt this extreme approach.

Several moderate approaches have been developed to meet the difficulties raised by these two extreme approaches. The core of the moderate approaches lies in the creation of probable consequence liability, that is, criminal liability for the unplanned offense, whose commission is the *probable consequence* of the planned original offense. Probable consequence refers both to mental probability from the point of view of the party participating in the offense, and to the factual consequence resulting from the planned offense. Thus, probable consequence liability generally requires two principal conditions to impose criminal liability for the unplanned offense:

1. a factual condition—the unplanned offense should be the consequence of the planned offense;
2. a mental condition—the unplanned offense should be probable (foreseeable by the relevant party) as a consequence of the commission of the planned offense.

The factual condition ("consequence") requires the incidental occurrence of the unplanned offense in relation to the planned one. There should be a factual causal connection between the planned and the unplanned offense. For example, A and B conspire to rob a bank, and execute their plan. During the robbery, B shoots the guard to death, an act that is incidental to the robbery and to B's role in it. Had it not been for the committed robbery, no homicide would have occurred. Therefore, the homicide is the factual consequence of the robbery, and it was committed incidentally to the robbery.

An incidental offense is one that was not part of the criminal plan, and the parties did not conspire to commit it. If the offense is part of the criminal plan, the probable consequence liability is irrelevant, and the general rules of complicity apply to the parties to the offense. Unplanned offenses fall short of this requirement and create an under-criminalization problem. The probable consequence liability is an attempt to address this difficulty by expanding the criminal liability for unplanned offenses despite the fact that they are unplanned.

The unplanned offense may be a different offense from the planned one, but not necessarily. The offense may also be an additional, identical offense. For example, A and B conspire to rob a bank by breaking into one of its safes. A is intended to break into the safe and B to watch the guard. They execute their plan, but in addition, B shoots and kills the guard, and A breaks into yet another safe. The unplanned homicide is a different offense from the planned robbery. The unplanned robbery is identical to the planned robbery. Both unplanned offenses are incidental consequences of the planned offense, although one is different from the planned offense and the other is identical to it. The planned offense serves as the causal background for both unplanned offenses, as they incidentally derive from it.[109]

The mental condition ("probable") requires that the occurrence of the unplanned offense be probable in the eyes of the relevant party, meaning that it could have been foreseen and reasonably predicted. Some legal systems prefer to examine the actual and subjective foreseeability (the party has actually and subjectively foreseen the occurrence of the unplanned offense), whereas others prefer to evaluate the ability to foresee through an objective standard of reasonability (the party has not actually foreseen the occurrence of the unplanned offense, but any reasonable person in his state could have). Actual foreseeability parallels the subjective *mens rea*, whereas objective foreseeability parallels objective negligence.

For example, A and B conspire to rob a bank. A is intended to break into the safe and B to watch the guard. They execute the plan, and B shoots and kills the guard while A breaks into the safe. In some legal systems, A is criminally liable for the killing only if A had actually foreseen the homicide, and in others, if a reasonable person could have foreseen the forthcoming homicide in these circumstances. Consequently, if the relevant accomplice did not actually foresee the unplanned offense, or any reasonable person in the same condition could not have foreseen it, he or she is not criminally liable for the unplanned offense.

This type of approach is considered moderate because it provides answers to the social endangerment problem, and at the same time it has a positive relation with the factual and mental elements of criminal liability. The factual and mental conditions are the starting terms and minimal requirements for the imposition of criminal liability for the unplanned offense.

Legal systems differ on the legal consequences of probable consequence liability. The main factor in these differences is the mental condition. Some legal systems require negligence, but others require *mens rea*, and the consequences may be both legally and socially different. Moderate approaches that are close to the extreme approach, which holds that all accomplices are criminally liable for the unplanned offense (*versanti in re illicita imputantur omnia quae sequuntur ex delicto*), impose full criminal liability for the unplanned offense if both factual and mental conditions are met. According to these approaches, the party is criminally liable for unplanned *mens rea* offenses even if he or she may have been merely negligent.

More lenient moderate approaches do not impose *full* criminal liability on all the parties for the unplanned offense. These approaches can show more leniency with respect to the mental element, and match the actual mental element of the party to the type of offense. Thus, the negligent party in the unplanned offense is criminally liable for a negligence offense, whereas the party who is aware is criminally liable for a *mens rea* offense.[110] For example, A, B, and C plan to commit robbery. The robbery is executed, and C shoots and kills the guard. A foresaw this, but B did not, although a reasonable person would have foreseen this outcome under the circumstances.

All three are criminally liable for robbery as joint perpetrators. C is criminally liable for murder, which is a *mens rea* offense. A, who acted under *mens rea*, is criminally liable for manslaughter or murder, both *mens rea* offenses. But B was negligent regarding the homicide, and is therefore criminally liable for negligent homicide. Negligent offenders are criminally liable for no more than negligence offenses, whereas other offenders, who meet *mens rea* requirements, are criminally liable for *mens rea* offenses.

American criminal law imposes full criminal liability for unplanned offense equally on all parties of the planned offense[111] as long as the unplanned offense is the probable consequence of the planned one.[112] Appropriate legislation has been enacted to accept the probable consequence liability, and has been considered constitutionally valid.[113] Moreover, in the context of homicide, American law incriminates incidental unplanned homicide committed in the course of the commission of another planned offense as murder, even if the mental element of the parties was not adequate for murder.[114]

By comparison, European-Continental legal systems and English common law impose criminal liability for the unplanned offense equally on all parties of the planned offense.[115] The English[116] and European-Continental moderate approaches are closer to the first extreme approach.[117]

For the applicability of the probable consequence liability to human-AI offenses, we must distinguish between two types of cases, as shown in our opening examples of computer programmers at the beginning of this section. The first type refers to cases in which the programmer designed the AI system to commit a certain offense, but the system exceeded the programmer's plan quantitatively (and committed more offenses of the same type), qualitatively (committed more offenses of different types), or in both respects. The second type refers to cases in which the programmer did not design the AI system to commit any offense, but the system committed an offense nevertheless.

In the first type of cases, criminal liability is divided into the liability for the planned offense and for the unplanned one. If the programmer designed the system to commit a certain offense, this is perpetration-through-another of that offense, at most. The programmer instructed the system what it should do, therefore he used the system instrumentally for the commission of the offense. In this case, the programmer is alone responsible for the offense, as already discussed.[118] For this particular criminal liability, there is no difference between an AI system, some other computerized system, a screwdriver, and any innocent human agent.

In the case of the exceeded offenses, a different approach is required. If the AI system is strong, and is capable of computing the commission of an additional offense, the AI system is considered criminally liable for that offense according to the standard rules of criminal liability, as described earlier.[119] This completes the criminal liability of the AI system, and the criminal liability of the programmer is determined according to the probable consequence liability already described. Thus, if from the programmer's point of view the additional offense was a probable consequence of the planned offense, then criminal liability is imposed on the programmer for the unplanned offense in addition to the criminal liability for the planned offense.

If the AI system is not considered strong, and is not capable of computing the commission of the additional offense, the AI system cannot be considered criminally liable for the additional offense. Under these conditions, the AI system is considered as innocent agent. The criminal liability for the additional offense remains the programmer's alone, on the same basis of probable consequence liability already described. Thus, if the additional offense was a probable consequence of the planned offense from the programmer's point of view, the programmer is criminally liable for the unplanned offense in addition to his criminal liability for the planned offense.

In the second type of cases, the programmer has no intention to commit any offense. From the programmer's point of view, the occurrence of the offense is not more than an unwilled accident. Because the programmer's initiative was not criminal, the probable consequence liability is inappropriate. The presence of a planned offense is crucial for the imposition of probable consequence liability, as discussed, because the probable consequence liability is intended to address unplanned developments of a planned delinquent event and serve as a deterrent against participation in delinquent activities.

When the programmer's starting point is not delinquent, and from his point of view the occurrence of the offense is accidental, then deterrence is inappropriate and irrelevant. Applying such a mechanism to deal with mistakes and accidents behind which there is no criminal intent would be disproportional. Thus, if the AI system, which actually committed the offense, is considered strong and capable of consolidating the requirements of the offense, it may be criminally liable for that offense as a direct perpetrator. If not, the AI system is not criminally liable for the offense.

The programmer is at most negligent. In this type of case, the programmer's criminal liability does not depend on the criminal liability of the AI system. Regardless of whether the AI system is criminally liable, the criminal liability of the programmer for the unplanned offense is examined separately. Because the programmer did not intend any offense to occur, the mental element of *mens rea* does not apply to him, and his criminal liability must be examined by the standards of negligence, so that he would be criminally liable for negligence offenses at most. The standard of negligence is discussed in the next chapter, where we discuss the criminal liability of AI entities for negligence offenses.

CLOSING THE OPENING EXAMPLE: INTENTIONAL KILLING ROBOT

We opened this chapter with the example of a killing robot. We can now go back and analyze the example based on the insights gained from this chapter. There are two main actors whose criminal liability must be evaluated in this homicide: the robot and its programmer. To examine the robot's criminal liability, we must first assess its technological capabilities. If the AI robot has the technological capabilities of meeting the requirements of *mens rea* offenses, it becomes a *machina sapiens criminalis* in this context. If such capabilities are present, the robot's criminal liability in this case should be examined as if it were human.

Homicide (*mens rea*) offenses can be murder or manslaughter. Their factual element requirement is identical, and it includes causing the death of a human. Because the robot has physically caused the worker's death, the factual element requirement appears to be met. The mental element requirement of murder requires intent (awareness and intent), whereas that of manslaughter requires recklessness (awareness and recklessness). At this point, careful analysis is needed. If, after analyzing the robot's record of computations, one of these requirements is met, it forms the criminal liability of the robot for the given homicide offense. If not, negligence may be examined, as discussed in the next chapter, and the robot's criminal liability is for negligent homicide. But if the robot lacks the technological capabilities to meet the requirements of either *mens rea* or negligence offenses, it does not form a *machina sapiens criminalis* in this context, and it is not criminally liable for the homicide. The only criminal liability for the homicide would be the programmer's and that of other related parties (users, the factory, and so on). Here, *programmer* refers not only to the specific person or persons who actually programmed the system, but to all related persons, including the company that manufactured the robot.

The programmer's criminal liability is determined based on his role in bringing about the homicide. If the programmer instrumentally used the robot to kill the worker by designing it to operate in this way, he is a perpetrator-through-another of the homicide. This is also the case if the robot has no advanced AI capabilities for meeting the requirements of the offense. In these cases, the programmer and the related parties are alone criminally liable for the homicide, and their exact criminal liability for the homicide, whether murder or manslaughter, is determined by the manner in which they met the mental element requirements of these offenses.

If the programmer designed the robot to commit an offense (any offense) other than the homicide of the worker, and the AI system of the robot continued the delinquent activity through an unplanned offense (the homicide of the worker), then the programmer's criminal liability for the homicide is determined based on probable consequence liability. The robot's criminal liability in this case is examined according to its capability to meet the requirements of the given offense. If the programmer did not design the robot to commit any offense, but the unplanned homicide occurred, then the robot's criminal liability is examined based on its capability to meet the requirements for homicide, and the programmer's criminal liability is examined according to the standards of negligence, as we will discuss in the next chapter.

3

AI CRIMINAL LIABILITY FOR NEGLIGENCE OFFENSES

One of the common applications of AI expert systems is medical,[1] generally used for more accurate diagnosis. Patient symptoms are entered into the system using visual scanners or other means to capture verbal data, after which the expert system analyzes the factual data and suggests a diagnosis to the medical staff. Based on machine learning (inductive analysis of cases and generalization), the expert system gains its "experience" and learns how to distinguish among various types of symptoms, and also to ignore unimportant symptoms. The longer the system is active, the more accurate is its diagnosis.

This method means, among other things, that identical cases brought to the system for diagnosis at different times might be diagnosed differently in light of the experience gained by the system between diagnoses. The same is true of human physicians. The case of one-year-old baby was brought to an expert AI system for diagnosis.[2] The baby suffered from high fever (104 degrees F), dehydration, and general weakness. The system diagnosed influenza, and recommended using analgesics and an infusion of fluids. The medical staff acted accordingly, but after five hours the baby died. The autopsy revealed that the baby died of a severe bacterial infection; if treated with the appropriate antibiotics, the baby could have survived. The legal question is, who is to be held criminally liable for the baby's death?

Negligence offenses are in common use in most countries worldwide in a context of professional malpractice. To impose criminal liability in these offenses, both factual and mental element requirements must be met by the apparent offender. Consequently, the question of the criminal liability of AI entities for negligence offenses is determined by whether they are capable of meeting these requirements accumulatively. But even if AI entities have such capability, their liability does not function as an exemption for humans from their own criminal liability, if humans were involved in the commission of the offense.

To explore these issues, we examine the capability of AI entities to fulfill both the factual and mental element requirements of negligence offenses, as well as the criminal liability of humans for AI negligence offenses.

3.1. THE FACTUAL ELEMENT REQUIREMENT

The factual element requirement structure (*actus reus*) is identical for all types of offenses: intentional, negligence, and strict liability, as noted earlier.[3] This structure contains one mandatory component (conduct) for all offenses, and two optional components (circumstances and results) for some offenses. The capability of machines to meet the factual element requirement considering this structure has already been discussed,[4] and the conclusions of the discussion are same within the context of negligence offenses.

3.2. THE MENTAL ELEMENT REQUIREMENT

Imposition of criminal liability for negligence offenses requires meeting both factual and mental element requirements. The mental element requirement of negligence offenses is negligence. If AI technology is capable of meeting the negligence requirement, it is possible and feasible to impose on such technology criminal liability for negligence offenses. First, we explore the structure of the negligence requirement, and then the fulfillment of the requirement by AI technology.

3.2.1. Structure of Negligence Requirement

Negligence has been recognized as a behavioral standard since ancient times. It was already mentioned in the Laws of Eshnunna in the twentieth century BC,[5] in Roman law,[6] in Canonic law, and in the early English common law.[7] But in these legal systems, negligence was treated as a behavioral standard rather than as a type of mental element in criminal law. The behavioral standard included dangerous behavior that fails to take into account all relevant considerations related to the individual's given action. Only since the seventeenth century has negligence been considered a type of mental element in criminal law.

In 1664, the English court ruled that negligence was not adequate to convict a defendant for manslaughter, and that at least recklessness was required.[8] This ruling created negligence as a type of mental element in criminal law. During the nineteenth century, negligence was treated as an exception for the general requirement of *mens rea*.[9] Accordingly, it was required explicitly and construed strictly. The offense needed to require negligence explicitly, in its definition, if it were to be adequate for imposing criminal liability. In the nineteenth century, negligence offenses were still quite rare.

As transportation developed, especially with the rise of automobiles, negligence started to be used more frequently in criminal law. Deaths caused by traffic accidents became more common, and manslaughter was not appropriate for these cases. A lower level of homicide was required, and the solution was found in negligent homicide.[10] But when negligence came into common use, it caused confusion. Negligence was interpreted as requiring unreasonable conduct, and this resulted in its being confused with recklessness of the lower level (rashness), which required taking unreasonable risk. That confusion produced such unnecessary terms as "gross negligence" and "wicked negligence."[11]

Many misleading rulings were issued on this basis in English law,[12] until the House of Lords clarified the distinction in 2003.[13] American law developed negligence as a mental element in criminal law, parallel to the English common law and inspired by it.[14] Negligence was accepted as an exception to *mens rea* during the nineteenth century, but more accurately than in English law.[15] The main distinction between recklessness and negligence involves the cognitive aspect of recklessness. Whereas recklessness requires the cognitive aspect of awareness, as part of the *mens rea* requirement, negligence does not.[16]

Both reckless and negligent offenders are required to take unreasonable risks. However, the reckless offender is required to be aware of the factual element components, but the negligent offender is not required to be so aware.[17] Negligence functions as an omission of awareness, and it creates a social standard of conduct. An individual is required to take only reasonable risks, measured objectively through the perspective of an abstract reasonable person.[18] The reasonable person is aware of his or her factual behavior, and takes only reasonable risks.[19] Naturally, it is the court that determines how reasonable a risk is, and it does so retrospectively in relation to a particular case.

The modern development of negligence in American law is reflected in the American Law Institute's Model Penal Code, inspired by modern European-Continental understandings of negligence. Accordingly, negligence is a type of mental element in criminal law, and as such it relates to the factual element components. Negligence requires lack of awareness of the factual element components, whereas a reasonable person could and should have been aware of them, and requires taking unreasonable risk regarding the results of the offense. This development has been embraced by American courts.[20]

How, then, can negligence function as a mental element in criminal law if the offender is not even required to be aware of his or her conduct? Some scholars have asked to exclude negligence from criminal law and leave it for tort law or other civil proceedings.[21] But the justification of negligence as mental element focuses on its function as an omission of awareness. Exactly the same way that act and omission are considered identical for the imposition of criminal liability, as discussed earlier in relation to the factual element,[22] so can both awareness and omission of awareness be considered as the basis for meeting the mental element requirement.

Following the analogy with the factual element, negligence does not parallel inaction, but parallels omission. If a person was simply not aware of the factual element components, and nothing more, she is not considered negligent, but innocent. Omitting to be aware means that the person was not aware although a reasonable person could and should have been; that is, the individual is considered to not be using her existing capabilities of forming awareness. If the individual was not aware but was capable of being aware, she also had the duty to be aware (*non scire quod scire debemus et possumus culpa est*).

Negligence does not incriminate individuals who are incapable of forming awareness, but only those who failed to use their existing capabilities to do so. Negligence does not incriminate the blind person for not seeing, but only people who have the ability to see but failed to use this ability. Wrong decisions are a common part of daily human life, so negligence does not apply to situations of this type. Taking risks is also part of human life, and therefore negligence is not intended to inhibit it, and indeed society encourages taking risks in many situations. Negligence applies to taking *unreasonable* risks.

If people were to take no risks at all, then human development would stop in its tracks. If we had taken no risks, we would have been staring at the burning branch after it was struck by lightning, afraid to reach out, grab it, and use it for our needs. Society constantly pushes people to take risks, but reasonable ones. The question is, how can we identify the unreasonable risks and distinguish them from the reasonable ones, which are legitimate?

For example, scientists propose to develop an advanced device that would significantly ease our daily lives. It is comfortable, fast, elegant, and accessible, but using it may cause the deaths of about 30,000 people each year in the United States alone. Would developing such a device be considered a reasonable risk?

Thirty-thousand victims a year appears to be an enormous number, which makes using the device completely unreasonable. But using the device in question, which we normally call a "car,"[23] is not considered unreasonable in most countries in the world today, although in the late nineteenth century it was. The same is true for trains, airplanes, ships, and many of our common means of transport and other everyday tools. The degree to which a risk is reasonable is relative to its nature, and is relatively determined with respect to time, place, society, culture, and other circumstances. Different courts in different countries find different types of people and behaviors to be reasonable within the context of negligence.

Whether a person was reasonable must be determined not only in the abstract, but also with reference to the relevant circumstances of the specific offender. For example, it is not appropriate to compare medical decisions of a physician with the behavior of an abstract reasonable person; such decisions should be compared with those of a reasonable physician of the same level of expertise and experience, acting in the same or similar clinical circumstances (emergency or other), having the same resources at his or her disposal, and so on. Such an approach would make the standard of reasonability more focused, but also subjective rather than purely objective. A detailed discussion of this process in relation to AI systems follows.[24]

Most negligence offenses are result offenses because society prefers to use negligence to protect against factual harms resulting from unreasonable risk taking. But negligence may be required for conduct offenses as well. The structure of negligence is described schematically in table 3.1.

The general structure of negligence has no volitive aspect, only a cognitive aspect. Because volition is supported by cognition, and negligence does not require awareness, negligence cannot require volitive components. The cognitive aspect of negligence consists of lack of awareness of all factual element components. The requirements of negligence in relation to conduct and circumstances are identical. Both require lack of awareness of the component (conduct/circumstances) despite the capability to form such awareness, when a reasonable person could and should have been aware of that component.

The reasonability in the components of negligence is examined with reference to the capability and duty of the offender to form awareness, although in practice, no awareness has been formed by him or her. With respect to the results, the negligence requirement requires lack of awareness of the possibility of the results occurring, despite the capability to form such awareness, when a reasonable person could and should have

Table 3.1 · General Structure of Negligence

FACTUAL ELEMENT COMPONENT	NEGLIGENCE COMPONENT
Conduct	Lack of awareness of conduct despite the capability to form such awareness, when a reasonable person could and should have been aware of that conduct.
Circumstances	Lack of awareness of circumstances despite the capability to form such awareness, when a reasonable person could and should have been aware of the circumstances.
Results	Lack of awareness of the possibility of the occurence of the results despite the capability to form such awareness, when a reasonable person could and should have been aware of that possibility as an unreasonable risk.

been aware of that possibility as an unreasonable risk. For this component, the reasonability focuses on identifying the occurrence of the result as a possibility involving unreasonable risk. In other words, under the given circumstances, the offender took unreasonable risks in the event at hand.

The modern structure of negligence is consistent with the concept of criminal law as a matrix of minimal requirements, described earlier.[25] It contains both internal and external aspects. Inwardly, negligence is the minimal mental element requirement for each of the factual element components. Consequently, if negligence is proven in relation to circumstances and results, but awareness is proven in relation to the conduct, this satisfies the requirement of negligence. This means that for each of the factual element components, at least negligence is required, but not exclusively. Outwardly, the mental element requirement of negligence offenses is satisfied through at least negligence, but not exclusively. This means that criminal liability for negligence offenses may be imposed by proving *mens rea* as well as negligence.

Because negligence is still considered an exception to the general requirement of *mens rea*, negligence has been required as the mental element of relatively light offenses. In some legal systems, negligence has even been restricted *ex ante* to light offenses.[26] This general structure of negligence is a template that includes terms from the mental terminology (for example, reasonability). To explore whether AI entities are capable of fulfilling the negligence requirement of given offenses, we must examine these terms.

3.2.2. Fulfillment of the Negligence Requirement: Is Objectivity Subjective?

Sonja: What prevents you from murdering somebody?
Boris: Murder's immoral.
Sonja: Immorality is subjective.
Boris: Yes, but subjectivity is objective.
—Diane Keaton and Woody Allen as Sonja and Boris Grushenko in *Love and Death*, United Artists, 1975

The core of the negligence template in relation to the factual element components is expressed by lack of awareness of the factual component despite the capability to form such awareness, when a reasonable person could and should have been aware of that component. Lack of awareness is naturally the opposite of awareness, which is required by *mens rea*, as discussed earlier.[27] Consequently, for a person to be considered aware of certain factual data, two accumulative conditions are required: (1) absorbing the factual data through the senses; and (2) creating a relevant general image with regard to this data in the brain. If either of these conditions is missing, the person is not considered to be aware.

Lack of awareness may be achieved by the absence of at least one of these two conditions. If the factual data has not been absorbed by the senses, or if it was absorbed but no relevant general image has been created, then the situation is considered one of lack of awareness to that factual data. Awareness is a binary situation in the context of criminal law; therefore, no partial awareness is recognized. If portions of the awareness process are present but the process has not been accomplished in full, the person is considered to be unaware of the relevant factual data. This is true for both human and AI offenders.

For lack of awareness to be considered omission of awareness, it must occur despite the capability of the offender to form awareness, when a reasonable person could and should have been aware. These are two separate conditions: (1) possessing the cognitive capabilities needed to consolidate awareness; and (2) a reasonable person could and should have been aware of the factual data. The first condition has to do with the offender's physical capabilities. If the offender lacks these capabilities, regardless of the offense, no criminal liability for negligence may be imposed on him, just as we cannot punish the blind for not seeing.

Therefore, negligent offenders are only those who possess the capability of forming awareness. That is true for both human and AI offenders. Therefore, to impose criminal liability for a negligence offense on an AI entity, the system must possess the capability of forming awareness. An AI system that lacks such capability cannot be held liable for an offense of negligence, and naturally it cannot be held liable for *mens rea* offenses either. These capabilities must be proven based on the general features of the AI system, regardless of the case at hand.

If we have established that the offender was not aware of the relevant factual data, and that he has the capability of being aware, for the offense to become negligence we must still prove that a reasonable person could and should have been aware of the factual data under similar circumstances. The "reasonable person" is a mental construct established for the purpose of comparison. Although in some other legal spheres, the reasonable person refers to a standard higher than that of the average person, in criminal law it refers to the average person.[28] Different societies and cultures, at different times and in different places, assign different content and qualities to the reasonable person.

The reasonable person is expected to reflect the existing situation in the given society, and not be used by the courts to change the current situation. The standard relates to the cognitive processes that should have occurred. The reasonable person is assessed based on two paths of cognitive activity: he or she should (1) take into account all relevant considerations, and (2) assign the proper weight to these considerations. Therefore, ignorance of relevant considerations is considered to be unreasonable, and after all relevant considerations have been taken into account, they must also be properly weighed for the action to be considered reasonable.

In general, the relevance of the considerations and their proper weight must be determined by the court *ex post*. The complexity and variety of common life situations makes it impossible to characterize a general type of reasonable person. Such a type would be purely objective and at the same time unrealistic and irrelevant to too many real-life situations. Therefore, the objective standard for the reasonable person must be set with some degree of subjectivity if it is to match any particular case. The subjective setting is the link between the individual offender and the objective and general standard of a reasonable person.

Thus, the reasonable person must be not only assessed by the standards of a general abstract person, but adapted to the specific circumstances of the individual offender. We expect different people to behave differently. Experienced attorneys act differently from inexperienced ones, even in the same situations. The same is true for pilots, drivers, physicians, police officers, and indeed, for all of us. Moreover, the same person acts differently under differing circumstances, weighing considerations differently. A soldier under enemy attack, in an emergency, in a life-threatening situation, acts differently than during routine training.

Therefore, the reasonable person relevant for a given case is not a general standard of objectivism, but it includes subjective reflections of the individual offender.[29] A surgeon with ten years of experience, acting in an emergency situation and using the limited resources available in the surgery, would be compared with a reasonable surgeon having the same attributes. The attributes relevant for that comparison, as well as their content and effect, are determined by the court, which may resort to the expert opinion of professionals for this purpose.

The reasonable person forms a sphere of reasonability that contains all types of reasonable behaviors in the given situation. The assumption is that there are several reasonable ways to behave in these situations, and only deviations from the sphere of reasonability can form negligence. When the factual data relate to the results of the conduct, the possibility of the occurrence of the results is considered an unreasonable risk. In this context, taking unreasonable risks is considered acting outside the sphere of reasonability with respect to the results.[30]

Reasonable and unreasonable risks are assessed in the same way as reasonable and unreasonable persons, as just described. For a risk to be considered reasonable, the individual must take into account all relevant considerations and assign the proper weight to them. Based on these considerations, if taking the risk is one of the available options, then the risk is considered reasonable; if not, it is considered unreasonable. To meet the negligence requirement, an AI entity must make unreasonable decisions. The ultimate question is whether a machine can be reasonable, or perhaps whether a machine can be unreasonable.

From an analytical point of view, the reasonability of a machine is not different from that of a human. Both must take into account the relevant considerations and assign the proper weight to them. This can easily be a matter of calculation. The relevant considerations are not more than terms in the equation, and their proper weight is the combinations of these terms. The equation may be constant if so programmed, but machine-learning features can change that. Machine learning is a process of generalization by induction from many specific cases, as we have already discussed.[31]

Machine learning enables the AI system to change the equation from time to time.

Indeed, effective machine learning should cause changes in the equation almost every time a given case is analyzed. This is what happens to our image of the world as our life experience becomes richer and broadens. Without machine learning, the equation remains constant and the system becomes ineffective. Expert systems that are not capable of machine learning are not different from human experts who insist on not updating their own knowledge. Machine learning is essential for the AI system to continue developing.

When the AI system is activated for the first time, the equation and its terms are programmed by human experts who determine what is a reasonable course of conduct in individual cases. After analyzing a few cases, the system begins to identify exceptions, wider and narrower definitions, new factors and new connections between existing ones, and so on. The AI system generalizes the knowledge absorbed from individual cases by reformulating the relevant equations. We use the term *equation* to describe the relevant algorithm, but it is not necessarily an equation in the mathematical sense.

Reformulating equations raises the possibility that the AI entity will make different decisions in the future than it made in the past. This process of induction is at the core of machine learning, and changes in the equation produce each time a different sphere of right decisions. For example, if on its first activation, a medical expert system is given a list of symptoms of both common cold and influenza, it produces a diagnosis based on those symptoms alone. But after being exposed to more cases, the system learns to take into account other symptoms that may be essential for distinguishing between the two conditions.

If the system is required to recommend medical treatments, these will differ for different diagnoses. At times, the expert system will not be "sure" because the symptoms can match more than one disease, and the system will assess the probabilities based on measurement and analysis of the various factors. For example, the expert system may determine that there is a probability of 47 percent that the patient has a common cold, and of 53 percent that she has influenza. Processing these probabilities may be the cause of negligence on the part of AI systems, and of mistaken conclusions, both sure and unsure. The system may be sure in the case of a mistaken conclusion, and can also assess probabilities mistakenly.

Mistakes may be caused by inauspicious changes of the equation, wrong factors being considered, ignorance of certain factors, or wrong weights assigned to some factors. These mistakes are the by-product of errors in the machine-learning process, or more precisely, *ex post* errors, which are considered errors only after the decision has been made and based on the consequences of that decision. Humans tend to learn empirically by trial and error. From the analytical point of view, in this context machine and human errors are of the same type. Understanding the error, its causes, and the ways to avoid it in the future are part of the learning process, for both humans and machines.

The question is, what is to be considered a reasonable decision in this context? The principal question is whether, in light of the system's initial state with respect to the basic factual data and the experience it has acquired through machine learning, a reasonable person could be aware of the relevant factual data. And the derivative question is, who is this reasonable person: a human or a machine? If we accept the concept that objectivity in negligence offenses is subjective to some degree, as already discussed, then the reasonable person must have attributes similar to those of the offender. Only then can the reasonability of the offender's decision be assessed without causing injustice.

Therefore, if the offender is capable of machine learning, so should be the reasonable person, according to this concept. It follows that in determining the reasonability of the AI entity's decisions, the reasonable person would be a reasonable AI system of the same type. It may be problematic for the human programmers, operators, and users to escape criminal liability and impose it on the mistaken machine. It would be convenient for the medical staff to use an expert system and follow its recommendations, then if the system is wrong, to have criminal liability imposed on it alone. But the legal situation is not that simple.

The decisions of assigning an AI system to its position, using it, following its recommendations, and so on are all subject to negligence offenses as well. The AI system is capable of meeting the negligence requirements, but this does not exempt other individuals involved in the situation from criminal liability. The very decision to use the AI system is itself subject to criminal liability. For example, if the decision was made with full awareness of particular mistakes that the system can make, and if these mistakes resulted in death, the human decision may lead to a charge of murder. But if there was no such awareness, although a reasonable person in a similar situation could and should have been aware of the risks, this may lead to a charge of negligent homicide.

Evaluation of the reasonable machine has to do with the machine-learning features of the system. The imposition of criminal liability in negligence offenses is based on analysis of the machine-learning process that led to the mistaken decision. The records maintained by the AI system itself are the basis for the analysis. But the reasonability of the decision-making process within the framework of machine learning may be based on expert opinion. This is how negligence is proved in court in the case of human offenders. It is not uncommon to prove or to refute the human negligence using expert opinions.

For example, when the medical expert system produces probabilities of 47 percent for common cold and 53 percent for influenza, a medical expert may explain to the court why these probabilities are reasonable or unreasonable for the case at hand, and a computer scientist may explain to the court the process of producing these probabilities based on the machine's learning process and existing database. Accordingly, the court must ask three questions:

1. Was the AI system unaware of the factual components?
2. Does the AI system have the general capability of consolidating awareness of the factual components?
3. Could a reasonable person have been aware of the factual components?

If the answer is affirmative to all three questions, and this is proven beyond any reasonable doubt, the AI system has met the requirements of the negligence offense. AI systems that are capable of forming awareness of *mens rea* offenses, as discussed earlier,[32] are not limited either technologically or legally in consolidating negligence for negligence offenses, because negligence requires a lower level of mental element than *mens rea* does. Negligence is therefore relevant to AI technology, and it is possible to prove it in court. But we must still answer the question of who is criminally liable for the commission of this type of offenses.

3.3. CRIMINALLY LIABLE ENTITIES FOR AI NEGLIGENCE OFFENSES

In general, imposition of criminal liability for negligence offenses requires meeting both the factual and mental element requirements of these offenses. Humans are involved in the creation, design, programming, and operation of AI technology and entities. Therefore, when the factual and mental elements of the offense are fulfilled by an AI entity, the question is, who is to be criminally liable for the offenses committed? The possible answers may be the AI entity, the related humans, or both. We will now discuss these answers.

3.3.1. AI Entity Liability

In negligence, as in *mens rea* offenses, when the offender meets both the factual and mental element requirements of an offense, criminal liability is imposed. As in *mens rea* offenses, the court is not expected to check whether the offender was "evil," "immoral," or so on, as noted earlier.[33] This is true for every type of offender: humans, corporations, and AI entities. Therefore, the justifications for the imposition of criminal liability on AI entities in *mens rea* offenses are relevant in the case of negligence as well. As long as these requirements are even narrowly met, criminal liability can be imposed.

Negligence offenses, however, differ from *mens rea* offenses in their social purpose. The pertinent question is whether their differing social purpose is relevant not only for humans and corporations, but also for AI entities. Initially, negligence offenses were not intended to deal with "evil" people, but with individuals who make mistakes. Therefore, the debate about villainy in criminal law is not relevant to negligence offenses, as it may be for *mens rea* offenses. In the context of negligence offenses, criminal law serves to help draw the boundaries of individual discretion.

Any person can make a wrong judgment, most of the time without contradicting the norms of criminal law. For example, we may choose the wrong spouse, car, employer, apartment, or even faith. But when mistaken use of our discretion leads to someone's death (negligent homicide), we do breach a norm of criminal law.[34]

As long as our exercise of wrong judgment does not contradict criminal law, society expects us to learn our lesson on our own, so that the next time we choose a car, house, employer, and so on, we will be more careful. This is how we gain life experience. In these cases, society takes the risk that we may not learn our lesson, but it does not intervene through criminal law. Yet when criminal offenses are committed, society does not take the risk of allowing the individual to be home-schooled, because the social harm is too serious to assume such risk.

In these cases, society intervenes by means of negligence offenses. The purpose of these offenses is to increase the likelihood that the individual will learn the relevant lesson. Prospectively, it is assumed that after the lesson has been taught, the probability of committing the same offense again is much lower. In this way, society educates physicians to be more careful during surgery, employers to be more assiduous in protecting their employees' lives and welfare, construction companies to be more scrupulous in observing safety regulations, factories to create less pollution, and so on. Human and corporate offenders are expected to learn their lessons through the intermediary of criminal law. Should this type of education be relevant to AI entities as well?

The answer is affirmative. For the educational purpose of the criminal negligence offense, it is true that there is not much use in the imposition of criminal liability unless the offender has the ability to learn. If we want the offender to learn a lesson after making a mistake, we must assume that the offender is capable of learning. If such capability is present and exercised, then criminal liability for negligence offenses is necessary. But if no such capability is exercised, then there is no prospective value in imposing criminal liability for negligence because the results would be the same, regardless of whether criminal liability is imposed.

For AI systems, which are equipped with the capabilities necessary for machine learning, criminal liability for negligence offenses is necessary. As for humans, negligence offenses can draw the boundaries of discretion of AI systems. Humans, corporations, and AI systems alike are expected to learn from their mistakes and improve their decisions prospectively. When the mistakes fall under its purview, criminal law intervenes in shaping the decision maker's discretion. For AI systems, criminal liability for negligence offenses is a chance to reconsider the decision-making process in light of external limitations dictated by criminal law.

We have learned over the years that improving the human process of decision making requires criminal liability for negligence, and the same logic applies to AI systems equipped with machine-learning functions. Although AI systems can be reprogrammed, their precious experience, gained through machine learning, would be lost. AI systems can learn the limits of their discretion on their own, but this is true for humans as well (and we must still impose criminal liability on humans for negligence offenses).

In sum, if AI entities are capable of fulfilling both the factual and mental element requirements of criminal liability for negligence offenses, and if the rationale for the imposition of criminal liability for these offenses is relevant to both humans and AI systems, there is no reason to avoid imposing criminal liability in these cases. But this is not the only type of AI involvement in criminal liability for negligence offenses.

3.3.2. Human Liability: Perpetration-through-Another and Semi-Innocent Agents

As described earlier, the most common way to deal with the instrumental use of individuals in the commission of offenses is the general form of perpetration-through-another.[35] To impose criminal liability for perpetration of an offense through another, it is necessary to prove awareness of such instrumental use. Therefore, perpetration-through-another is applicable only in the case of *mens rea* offenses. In most cases, the person being instrumentally used by the perpetrator is considered an innocent agent, and no criminal liability is imposed on him or her. The analysis of perpetration-through-another in the context of *mens rea* offenses has already been discussed.

Nevertheless, a person instrumentally used can also be considered a semi-innocent agent who is criminally liable for negligence, although at the same time, the perpetrator is criminally liable for a *mens rea* offense. This is the case when negligence may be relevant for the perpetration-through-another, and that completes the discussion toward it. For example, a nurse in an operating room realizes that a person who had attacked her in the past is about to be operated on, and she decides that he deserves to die. She infects the surgical instruments with lethal bacteria, telling the surgeon that the instruments have been sterilized.

A few hours after the surgery, the patient dies as a result of an infection. Legal analysis of the case suggests that the nurse is the perpetrator-through-another of murder, having instrumentally used the surgeon to commit the murder. The surgeon's criminal liability in this case depends on his mental state. If he is an innocent agent, he is exempt from criminal liability. But if he has the legal duty to make sure that the instruments have been sterilized, he is not an entirely innocent agent because he failed to fulfill his legal duties.

At the same time, because he was not aware of the infected instruments, this is a case of negligence. When the agent is not aware of crucial elements of the offense, but a reasonable person in the same situation could and should have been aware, the agent is negligent and is called a semi-innocent agent.[36] Thus, when a person instrumentally uses another person who is negligent regarding the commission of the offense, this is perpetration-through-another, but both persons are criminally liable: the perpetrator for a *mens rea* offense (for example, murder) and the other person for a negligence offense (for example, negligent homicide). Given that AI systems are capable of forming negligence as a mental element, the question is whether they can function as semi-innocent agents.

A case in which an AI system can be a semi-innocent agent is one in which the perpetrator (human, corporation, or AI entity) instrumentally uses an AI system for the commission of an offense, and although it was used instrumentally, the AI system was negligent with regard to committing the offense. Only AI systems that are capable of fulfilling the mental element requirement of negligence offenses can be considered as semi-innocent agents and can function as such. But not in every case in which the AI system has the capability for negligence would it automatically function as a semi-innocent agent. The capability for negligence is necessary for functioning as a semi-innocent agent, but it is not sufficient.

The semi-innocent agent, whether human, corporation, or machine, must be examined *ad hoc* in any given case. Only if the agent was negligent regarding the commission of the offense can it be considered a semi-innocent agent. Thus, if the instrumentally used AI system did not consolidate awareness of the relevant factual data but had the capability to do so, and a reasonable person would have consolidated such awareness, the AI system is considered a semi-innocent agent in the context of a perpetration-through-another (itself).

The perpetrator's criminal liability is not affected by the agent's criminal liability, if any. The criminal liability of the perpetrator-through-another is for the *mens rea* offense, whether the instrumentally used AI system has no criminal liability (that is, is an innocent agent or lacks the relevant capabilities) or is criminally liable for negligence (that is, is a semi-innocent agent). Thus, using the legal construction of perpetration-through-another based on the instrumental use of an AI entity does not affect the status of the perpetrator as a *mens rea* offender, regardless of the AI system's criminal liability.

By the same token, the agent's criminal liability in these cases is not directly affected by the perpetrator's criminal liability. If the AI system was negligent (that is, it fulfilled both factual and mental element requirements of the negligence offense), then criminal liability for negligence is imposed on it, and the system is classified as a semi-innocent agent in a perpetration-through-another. If the AI system was not negligent because of its lack of capability or for any other reason, then no criminal liability is imposed on it, and the system is classified as an innocent agent in a perpetration-through-another.

In other words, if the AI system is neither an innocent agent nor a semi-innocent agent, this means that it has fully met the requirements of the *mens rea* offenses. This is not the case for perpetration-through-another, but for principal perpetration on the part of the AI system. If the AI system is capable of being criminally liable for *mens rea* offenses as a sole offender, then there is nothing to prevent it from committing the offense jointly, with other entities, humans, corporations, or other AI entities. Complicity by the AI systems requires at least *mens rea*, not negligence, because it requires awareness of the complicity and of the delinquent association. This situation is not substantively different from meeting the requirements of any other *mens rea* offense, as discussed earlier.[37]

3.3.3. Joint Human and AI Entity Liability: Probable Consequence Liability

Probable consequence liability deals with the commission of an unplanned offense, different from or additional to the main, planned offense by more than one perpetrator. The question in these cases involves the criminal liability of the other parties to the unplanned offense committed by one of the parties. We have already discussed probable consequence liability for an unplanned *mens rea* offense.[38] The relevant question here concerns probable consequence liability to an unplanned negligence offense committed by an AI system. Are the programmers, users, and other related persons criminally liable for unplanned negligence offense committed by an AI system?

This question winds up our two previous discussions regarding human criminal liability for negligence by AI entities in addition to or instead of the criminal liability of the AI system for that offense.[39] For example, a medical expert system is used to diagnose certain types of diseases by analyzing patient symptoms. The analysis is based on machine learning and inductive generalization of various cases. When the system fails to diagnose a case correctly, this results in the wrong treatment, leading eventually to the patient's death. Analysis of the system's activity reveals negligence by it, which meets both the factual and mental element requirements of negligent homicide. At this point a question arises about the programmer's criminal liability for the offense.

The programmer's criminal liability has nothing to do with the decision to use the AI system, to follow its diagnosis, and so on, but involves the initial programming of the system. If the programmer had programmed the system to kill patients and had instrumentally used it for this purpose, this would have been perpetration-through-another of murder, but this is not the case here. In this case, the probable consequence liability may be relevant for the programmer's criminal liability.

The mental condition for the probable consequence liability requires the unplanned offense to be "probable" from the point of view of the party that did not actually commit it. It is necessary for that party to have been able to foresee and reasonably predict the commission of the offense. Some legal systems prefer to examine actual and subjective foreseeability (the party has actually and subjectively foreseen the occurrence of the unplanned offense), whereas others prefer to evaluate the ability to foresee through an objective standard of reasonability (the party has not actually foreseen the occurrence of the unplanned offense, but any reasonable person in this situation could have).

Actual foreseeability parallels subjective *mens rea*, whereas objective foreseeability parallels objective negligence. Consequently, in legal systems that require objective foreseeability, the programmer should be at least negligent for the commission of the negligence offense by the AI system for the imposition of criminal liability for that offense. But in legal systems that require subjective foreseeability, the programmer must be at least aware of the possibility of the commission of the negligence offense by the AI system for imposition of criminal liability for that offense.

However, if neither subjective nor objective foreseeability of the commission of the offense apply to the programmer, the probable consequence liability is irrelevant. In this case, no criminal liability can be imposed on the programmer, and the criminal liability of the AI system for the negligence offense does not affect the programmer's liability.

CLOSING THE OPENING EXAMPLE: NEGLIGENT KILLING ROBOT

We opened this chapter with an example describing an expert system that killed through negligence. We can now return to this case and analyze it based on insights achieved in this chapter. There are four main characters whose criminal liability can be considered for this homicide: the AI system, the programmer, the hospital, and the user of the AI system. To examine the criminal liability of the AI system, we must examine first its technological capabilities. If the AI system has the technological capabilities to meet the requirements of negligence offenses, it forms a *machina sapiens criminalis* in this context.

If these capabilities are present, the AI system's criminal liability should be examined as if it were human. But if the AI system does not have the technological capabilities to meet the requirements of negligence offenses, it cannot be considered a *machina sapiens criminalis* and is not criminally liable for the negligent homicide. In the opening example, the medical expert system met the factual element requirement for negligent homicide because its diagnosis and recommendations set off a chain of events that ended with the baby's death.[40]

To meet the mental element requirement, it is necessary to prove negligence on the part of the system. To do so, it is required to show that the system has the capability to consolidate awareness, that it did not in practice form awareness of the relevant factors involved in the mistaken diagnosis, and that a reasonable person could have formed such awareness. The first condition can be proven based on the characteristics of the system and according to the definition of awareness in criminal law (perception by the senses of factual data and its understanding), as discussed earlier.[41] If the system has such capability, and if that capability has been activated, then the system formed awareness.

If the system had been aware of the bacterial infection and nevertheless produced an inappropriate diagnosis and recommendations, this would not have been negligence but *mens rea*, and the homicide offense would have been manslaughter or murder. Otherwise, the system meets the second condition for negligence. For the AI system to be criminally liable for negligent homicide, its mental state must be compared with that of a reasonable person. The subjective incarnation of the reasonable person in this case is a reasonable medical expert system. The question is whether the system took into consideration all the relevant factors required for proper diagnosis, and whether they were appropriately weighed, given the relevant circumstances of the case.

The diagnosis produced by the system did not necessarily have to be correct for the system to be exempt from liability for negligence. Not every incorrect diagnosis is considered negligent. At issue here is the system's judgment. If it is proven that the system's judgment was correct (the calculation took all factors into consideration and weighed them properly), then the system acted reasonably. Although the diagnosis did not correctly identify the cause of the symptoms, this is still not in the realm of negligence if the system's judgment was reasonable. Not every mistake is considered negligent. In this case, the court would have to examine circumstances carefully, with or without the help of medical and software experts.

If the court reaches the conclusion that a reasonable person would have been aware of the possibility that the symptoms were caused by a bacterial infection, or would have considered the probability to be low that influenza had caused the symptoms, then the system is considered negligent. Otherwise, the system is exempt from criminal liability for the death of the baby. Three other actors may be criminally liable in this case, in addition to (but not instead of) the AI system. The criminal liability of the AI system may affect that the liability of these actors, but not necessarily.

The programmer's criminal liability is determined by his role in the occurrence of the death. If the programmer instrumentally used the AI system to kill the baby by designing it in such a way as to produce erroneous medical diagnoses, the programmer is a perpetrator-through-another of the homicide. This would also be the case when the AI system had no advanced AI capabilities to meet the requirements of the offense, so that the programmer and the other related parties, but not the AI system, are the only ones criminally liable for the homicide. Their exact criminal liability for the homicide, whether murder or manslaughter, is determined by whether they meet the mental element of these offenses.

If the programmer has foreseen the consequence (the patient's death), or if any reasonable programmer could have foreseen it, and if the AI system is criminally liable for the homicide, the programmer may be criminally liable for negligent homicide through probable consequence liability, if the offense was unplanned. Generally, if the programmer did not intentionally design the system to commit any offense, and an unplanned homicide (murder, manslaughter, or negligent homicide) occurred, the system's criminal liability is examined based on its capabilities and on meeting the requirements of the relevant homicide, and the programmer's criminal liability is examined based on the standards of negligence. Here, *programmer* refers not only to the person or persons who actually programmed the system, but to all related persons, including the company that produced the system.

Two other actors who may be criminally liable are the users of the system (the medical staff) and the hospital. The hospital may be criminally liable for making the decision to use an AI system in real-life medical cases. To determine the liability of the hospital, it is necessary to examine the factual data about the system that was available and known to the hospital, including the probability that the system would make mistakes. If the decision was reasonable, the hospital is not negligent. Similarly, the medical staff must be examined to determine its criminal liability for homicide. The factual data about the system that was available and known to the staff is extremely important in evaluating the decision of the medical staff to follow the system's diagnosis and recommendations.

If the AI system was correct in the previous 1,000 cases, then the medical staff's decision to follow its recommendations may be reasonable. But it is up to the court to determine the sphere of reasonability in each case. Perhaps in emergency rooms a 10 percent rate of mistakes is reasonable, whereas in cosmetic surgery such a rate is far beyond reasonability. Thus, criminal liability of AI systems for negligence offenses is possible, and it does not necessarily reduce the criminal liability of other relevant entities.

4

AI CRIMINAL LIABILITY FOR STRICT LIABILITY OFFENSES

The operations carried out by drones are based on artificial intelligence technology. AI drones operate not only on the ground (unmanned vehicles), but also in the air and underwater.[1] For example, in 2012 the US Navy examined an operative drone that was slated to land on an aircraft carrier, relying on pinpoint CPR coordinates and advanced avionics. Computers on the carrier digitally transmit the relevant data (speed, crosswinds, and so on) to the drone, as it approaches for the landing. The drone not only knows how to land by itself, but it also knows the weapons it is carrying; if, when, and where it needs to refuel from an aerial tanker; and whether there is a nearby threat. It does its own calculations and decides what to do next.[2]

The same technology is implemented in other devices as well, with the possible result that these devices can commit traffic offenses, whether in the air or on the ground. For example, an unmanned vehicle is used for ground transportation on roads travelled by other, human-driven vehicles. When the unmanned vehicle arrives at an intersection, it analyzes the traffic situation and decides to proceed. The vehicle has a green light, and the lights in all other directions are red, but an ambulance arrives at high speed and crosses the intersection.

The vehicle hits the ambulance, causing severe injury to the passengers, and the patient in the ambulance dies as a result. Analysis of the patient's medical situation shows that the patient would have survived had the ambulance not been hit by the unmanned vehicle, so the collision is the direct factual cause of the patient's death. The legal question is, who is to be held criminally liable for the patient's death?

Strict liability offenses are in common use by most countries in a context of public order and welfare (for example, traffic offenses, labor safety offenses, or environmental offenses). For imposition of criminal liability in these offenses, both the factual and mental element requirements must be met by the apparent offender. Consequently, the question regarding the criminal liability of AI entities for strict liability offenses is whether they are capable of meeting these requirements accumulatively. Moreover, even if these entities have such capability, this does not serve to exempt humans from criminal liability if they were involved in the commission of the offense.

To explore these issues, we examine the capability of AI entities to meet both the factual and mental element requirements of strict liability offenses, and the criminal liability of human offenders for strict liability offenses committed by AI entities.

4.1. THE FACTUAL ELEMENT REQUIREMENT

The factual element requirement structure (*actus reus*) is identical for all types of offenses: intentional, negligence, and strict liability, as noted earlier.[3] This structure contains one mandatory component (conduct) for all offenses, and two optional components (circumstances and results) for some offenses. The capability of machines to meet the factual element requirement, considering this structure, has already been discussed,[4] and the conclusions of that discussion hold true for negligence offenses as well.

4.2. THE MENTAL ELEMENT REQUIREMENT

Imposition of criminal liability for strict liability offenses requires meeting both factual and mental elements requirements. The mental element requirement of strict liability offenses is strict liability or presumed negligence. If AI technology is capable of meeting the strict liability requirement, it is possible and feasible to impose criminal liability on it for strict liability offenses. We begin by exploring the structure of strict liability requirements, and then proceed to examine the way in which AI technology meets the requirement.

4.2.1. Structure of Strict Liability Requirement

Strict liability evolved from absolute liability and was accepted in criminal law as a form of mental element requirement. Beginning in the eighteenth century, English common law determined that some offenses require neither *mens rea* nor negligence. These offenses, called public welfare offenses, were inspired by tort law, which accepted absolute liability as legitimate.[5] Consequently, these became criminal offenses of absolute liability, and imposition of criminal liability for them required proof of the factual element alone.[6]

Absolute liability offenses were considered exceptional because they require no mental element. In some cases Parliament intervened and required a mental element,[7] and in other cases mental element requirements were added by court rulings.[8] By the mid-nineteenth century, English courts began to consider efficiency criteria as part of criminal law in various contexts, which gave rise to convictions on the basis of public inconvenience. Offenders were indicted and convicted despite the fact that no mental element was proven because of the public inconvenience caused by the commission of the offense.[9]

These convictions created, in practice, an upper threshold for negligence, a type of augmented negligence, which required individuals to follow strict guidelines and make sure they committed no offenses. This standard of behavior is higher than the one imposed by negligence, which requires only reasonable conduct. Strict liability offenses require more than reasonability, making sure that no offense is committed at all. These offenses show a clear preference of the public welfare over the strict justice for the potential offender. Because these offenses were not considered severe, they were expanded "for the good of all."[10]

This development was considered necessary because of the legal and social changes at the time of the first industrial revolution. For example, the increasing number of workers in the cities caused employers to degrade the workers' social conditions. Parliament intervened through social welfare legislation, and the efficient enforcement of this legislation was through absolute liability offenses.[11] It was immaterial whether the employer knew what the proper social conditions for workers were; employers had to make sure that no violation of these conditions occurred.[12] In the twentieth century, this type of criminal liability spread to other spheres of law as well, including traffic law.[13]

American criminal law accepted absolute liability as a basis for criminal liability in the mid-nineteenth century,[14] ignoring previous rulings that rejected it.[15] Acceptance was restricted to petty offenses, and violations were punished by fines, and not very severe ones at that. Similar acceptance occurred at about the same time in the European-Continental legal systems, so that absolute liability in criminal law became a global phenomenon.[16] In the meantime, the fault element in criminal law became much more important because of internal developments in criminal law, and *mens rea* became the major and dominant requirement for mental element in criminal law.

Therefore, criminal law was required to make changes in absolute liability in order to meet the modern understanding of fault, which triggered the move from absolute to strict liability. At the heart of the change was the move from absolute legal presumption (*praesumptio juris et de jure*) to relative legal presumption (*praesumptio juris tantum*), so that the offender had the opportunity to refute the criminal liability. The presumption was presumption of negligence, either refutable or not.[17] The move from absolute liability to strict liability eased the acceptance of the presumed negligence as yet another, third form of mental element in criminal law.

Since the broad acceptance of strict liability worldwide, legal systems have justified it both from the perspective of fault in criminal law[18] and constitutionally. The European Court of Human Rights justified using strict liability in criminal law in 1998,[19] and thus strict liability was considered not to be contradicting the presumption of innocence, protected by the 1950 European Human Rights Covenant.[20] This ruling has been embraced in Europe and in Britain.[21] The Supreme Court of the United States ruled consistently that strict liability does not contradict the US Constitution.[22] So did the supreme courts in various states.[23] But in the United States, the courts recommended to restrict the use of these offenses to the minimum necessary, and to prefer using *mens rea* or negligence offenses.

The strict liability construct in criminal law focuses on the relative negligence presumption and the ways to refute it. According to this presumption, if all components of the offense's factual element requirement are proven, it is presumed that the offender was at least negligent. Consequently, for the imposition of criminal liability in strict liability offenses, the prosecution does not have to prove the mental state of the defendant, only the fulfillment of the factual element requirements. The mental state of the offender is derived from the conduct. Up to this point, strict liability is similar to absolute liability.

But contrary to absolute liability, strict liability can be refuted by the defendant because it is based on a *relative* legal presumption. To refute strict liability, the offender must accumulatively meet two conditions: (1) no *mens rea* or negligence existed on the part of the offender, and (2) all reasonable measures to prevent the offense were taken.

The first condition refers to the mental state of the offender. According to the presumption, the commission of the factual element presumes that the offender was *at least* negligent. This means that the offender's mental state was one of negligence or of *mens rea*. The offender must therefore refute the conclusion of the presumption, and prove it incorrect in his case. To do so, the offender must prove that he was not aware to the relevant facts, and that no other reasonable person could have been aware of them under the same circumstances. This proof resembles refuting *mens rea* in *mens rea* offenses and negligence in negligence offenses.

However, strict liability offenses are not *mens rea* or negligence offenses, because refuting *mens rea* and negligence is not sufficient to prevent imposition of criminal liability. The social and behavioral purpose of these offenses is to *ensure* that individuals conduct themselves strictly according to rules, and commit no offenses. That should be proven as well. The offender must therefore prove that he or she took *all* reasonable measures to prevent the offense.[24] Note the difference between strict liability and negligence: to refute negligence, it is sufficient to prove that the offender acted reasonably, but to refute strict liability, it is necessary to prove that *all* reasonable measure were taken.[25]

To refute the negligence presumption of strict liability, the defendant must positively meet both conditions by a preponderance of the evidence, as in civil law cases. The defendant is not required to prove that these conditions have been met beyond any reasonable doubt, but in general, it is not sufficient to merely raise a reasonable doubt. The burden of proof is higher than the general burden of the defendant. The possibility for the offender to refute the presumption becomes part of the strict liability requirement because it relates to the offender's mental state. The structure of strict liability is described schematically in table 4.1.

The modern structure of strict liability is consistent with the concept of criminal law as a matrix of minimal requirements, described earlier,[26] and it has both internal and external aspects. Internally, strict liability is the minimal mental element requirement for each of the factual element components. Consequently, if strict liability is proven in relation to circumstances and results but negligence is proven in relation to conduct, that satisfies the requirement of strict liability. This means that for each of the factual element components, at least strict liability is required, but not exclusively strict liability. Externally, the mental element requirement of strict liability offenses is satisfied by at least strict liability, but not exclusively. This means that criminal liability for strict liability offenses may be imposed by proving *mens rea* and negligence as well as strict liability.

Because strict liability is still considered an exception to the general requirement of *mens rea*, it is an adequate mental element for relatively light offenses. In some legal systems, strict liability has been restricted *ex ante* or *ex post* to light offenses.[27] This general structure of strict liability is a template that contains terms derived from the mental terminology. To

Table 4.1 · General Structure of Strict Liability

FACTUAL ELEMENT COMPONENT	STRICT LIABILITY COMPONENT
Conduct	The conduct has been committed, and the offender fails to prove that neither *mens rea* nor negligence requirements for the conduct have been met, and that all reasonable measures to prevent the offense were taken.
Circumstances	The circumstances are present, and the offender fails to prove that neither *mens rea* nor negligence requirements for the conduct have been met, and that all reasonable measures to prevent the offense were taken.
Results	The results have occurred, and the offender fails to prove that neither *mens rea* nor negligence requirements for the conduct have been met, and that all reasonable measures to prevent the offense were taken.

explore whether AI entities are capable of meeting the strict liability requirement of given offenses, we must examine these terms.

4.2.2. Fulfillment of Strict Liability: Making the Factual Become Mental

To prove that the defendant met the requirement for strict liability, the prosecution may choose to prove only the factual element of an offense. According to the negligence presumption, the presence of the factual element indicates that the offender has been at least negligent. But as already noted,[28] the possibility of refuting that presumption by the defendant is an integral part of the substance and structure of strict liability. Therefore, strict liability may apply only to offenders who possess the mental capability of refuting the presumption, which has to do with the mental element.

The mental capability of refuting the presumption does not necessarily mean that the offender has proof of his or her innocence, or has convincing arguments to that effect, but only that the inner capabilities required to refute the negligence presumption are present. Refuting the presumption requires proof that the offender (1) was neither aware nor negligent regarding the factual element components, and (2) has taken all reasonable measures to prevent the commission of the offense. The required capabilities, therefore, are those needed to consolidate awareness and negligence, and those needed to act reasonably.

The question is whether AI systems possess such capabilities, and therefore whether criminal liability for strict liability offenses may be imposed on them. We examine these capabilities one by one. First is the capability to consolidate awareness and negligence. We have examined this capability of AI systems already in relation to *mens rea* and negligence.[29] AI systems that possess the relevant features discussed earlier do indeed have this capability. Consequently, all AI systems that can be indicted for *mens rea* and negligence offenses can also be indicted for strict liability offenses.

This makes sense. Analogously, if *mens rea* and negligence require higher mental capabilities than does strict liability, then the lower capability required for strict liability would be much easier to achieve than the higher capabilities needed for *mens rea* and negligence. Therefore, the AI offender is required to possess the same features, regardless of the type of offense. Because negligence requires the capability to consolidate awareness,[30] and because strict liability requires the capabilities to consolidate awareness and negligence, it is clear that all AI offenders are required to possess the capability of consolidating awareness.

So in this context, an indictable offender is one who has the *capability* of consolidating awareness, regardless of whether this capability has been realized and utilized, and regardless of the type of offense. An AI system that is indictable for strict liability offenses must have the mental features and capabilities needed to be indicted for *mens rea* or negligence offenses. This fact reveals the character of the AI offender, with respect to inner capabilities, as more or less uniform. Thus, the minimal inner features and capabilities of *machina sapiens criminalis* are uniform and can be defined accordingly.

The second required capability is reasonability. Refuting the negligence presumption requires proof of having taken all reasonable measures, as already noted, and the capability of acting reasonably. Without such capability, no AI system can take all reasonable measures. The capability for reasonability is required in negligence as well. This is the same capability as the one required for strict liability, and we have discussed it already in the context of negligence.[31] The fact that the same capability is required strengthens the argument just made that all inner capabilities of *machina sapiens criminalis* are uniform, regardless of the type of offense.

Although the capability is the same in both negligence and strict liability offenses, it operates differently in the cases of negligence and strict liability. For negligence offenses, it is required that the AI system identify the reasonable options and make a choice between them. The practical requirement is that its final choice be a reasonable option. For strict liability offenses, it is also required that the AI system identify the reasonable options, but in addition, it must choose to exercise all of them. The choice here is between reasonable and unreasonable, whereas in the case of negligence, the choice is among the reasonable options.

It is much easier for an AI system, under these conditions, to act reasonably in a strict liability context than in the context of negligence because in strict liability situations, fewer choices are required. Accordingly, in relation to the criminal liability of an AI system in strict liability offenses, the court must answer the following three questions:

1. Was the factual element of the offense fulfilled by the AI system?
2. Does the AI system have the general capability of consolidating awareness or negligence?
3. Does the AI system have the general capability of reasonability?

If the answer to all three questions is affirmative, and this is proven beyond any reasonable doubt, then the AI system has fulfilled the requirements of the strict liability offense, and the AI system is presumed to be at least negligent.

At this point, the defense has the opportunity to refute the negligence presumption through positive evidence. After the evidence is presented, the court must decide in two questions:

1. Has the AI system indeed formed *mens rea* or negligence regarding the factual element components of the strict liability offense?
2. Has the AI system not taken *all* reasonable measures to prevent the actual commission of the offense?

If the answer to even one of these two questions is affirmative, the negligence presumption has not been refuted, and criminal liability for the strict liability offense is imposed. Only if the answers to both questions are negative is the negligence presumption refuted, and no criminal liability for the strict liability offense can be imposed on the AI system. In general, AI systems that are capable of forming awareness and negligence, as discussed earlier,[32] have neither technological nor legal limitations in forming the inner requirements for strict liability offenses because strict liability requires a lower level of mental element than do *mens rea* or negligence. Given that strict liability is relevant to AI technology and that it is possible to prove it in court, then who is to be criminally liable for the commission of this type of offenses?

4.3. CRIMINALLY LIABLE ENTITIES FOR AI STRICT LIABILITY OFFENSES

In general, imposition of criminal liability for strict liability offenses requires meeting both factual and mental elements of these offenses. Humans are involved in the creation, design, programming, and operation of AI technology and entities. Consequently, when the factual and mental elements of the offense are met by the AI entity, the question is, who is to be criminally liable for the offenses committed? The possible answers are the AI entity, the humans, or both. We will now discuss these answers.

4.3.1. AI Entity Liability

In strict liability offenses, such as *mens rea* and negligence, when the offender fulfills both the factual and mental element requirements of the offense, criminal liability is imposed. As in the case of *mens rea* and negligence offenses, the court is not supposed to check whether the offender had been "evil" or "immoral," as already discussed.[33] This is true for all types of offenders, including humans, corporations, and AI entities. Therefore, the justifications used to impose criminal liability on AI entities in *mens rea* and negligence offenses are relevant for strict liability as well. As long as these requirements are narrowly met, criminal liability can be imposed.

But strict liability and negligence offenses are different from *mens rea* offenses in their social purpose, as we have discussed.[34] The relevant question is whether their different social purpose applies not only to humans and corporations, but to AI entities as well. From their inception, strict liability offenses were not designed to deal with evil people, but with individuals who did not make all possible efforts to prevent the commission of an offense. Therefore, the debate about evil in criminal law is not relevant to strict liability offenses in the same way that it may be relevant for *mens rea* offenses.

In this context, criminal law function is to educate and ensure that no offense is committed, and strict liability offenses serve to draw the boundaries of individual discretion. Any person can use wrong judgment and make some, but not all possible efforts to prevent the commission of an offense. Most of the time, this does not contradict the norms of criminal law. For example, we may take the risk of investing our money in doubtful stocks, and we may not take all possible precautions to prevent damage to our investments, but that does not breach any criminal law.

In some cases, however, wrong judgment contradicts a norm of criminal law that was designed to educate individuals to make sure that no offense is committed. As long as our wrong judgment and lack of effort to prevent offenses do not contradict the criminal law, society expects us to learn our lesson on our own. Next time we invest our money, we will examine the details of our investment more carefully. This is how we gain life experience. In this case, society takes a risk that we will not learn our lesson, but still does not intervene through criminal law.

But when criminal offenses are committed, society does not assume the risk of allowing individuals to learn their lessons on their own. In these cases the social harm is too grave to allow such risk, and society intervenes using both strict liability and negligence offenses. In the case of strict liability offenses, the purpose is to educate individuals to make every possible effort to prevent the occurrence of the offense. For example, drivers are expected (and educated) to drive carefully to prevent any possible moving violation.

The objective of the offense is to make sure that individuals learn how to behave extra carefully. Prospectively, it is assumed that after someone has been convicted of a strict liability offense, the probability for recommission of the offense is much lower. For example, society educates its drivers to drive extra carefully, and its employers to adhere to state regulations regarding the payment of wages, and so on. Human and corporate offenders are expected to learn how to behave carefully by means of the criminal law. Is this method of education relevant for AI entities as well?

The answer is affirmative. For the educational purpose of strict liability, there is not much use in imposing criminal liability unless the offender has the ability to learn and change his or her behavior accordingly. If we want to make offenders learn to behave very carefully, we must assume that they are capable of learning and implementing that knowledge. If such capabilities are present and exercised, it is necessary to impose criminal liability for strict liability offenses. But if no such capabilities are exercised, then imposing criminal liability is entirely unnecessary because no prospective value is expected, and the result of using or not using criminal liability for strict liability offenses would be the same.

For AI systems that are equipped with the relevant capabilities for machine learning, criminal liability for strict liability offenses is necessary if these capabilities are to be applied in the relevant situations in which there is an obligation to behave extra carefully. Exactly as for humans, strict liability offenses may trace the boundaries of discretion for AI systems. Humans, corporations, and AI systems alike are supposed to learn from their experience and improve their decisions prospectively, including the standards of carefulness.

When the absence of carefulness is part of criminal law, the law intervenes to shape the judgment exercised with respect to careful behavior. For the AI system, the criminal liability for strict liability offenses is an opportunity to reconsider the decision-making process in light of external limitations dictated by criminal law, which require extra-careful conduct and decision making. We have learned over many generations that the human decision-making process requires criminal liability for strict liability in order to be improved, and the same logic applies to AI systems using machine learning methods.

Naturally, AI systems can simply be reprogrammed each time, but this is different from the precious experience gained through machine learning. AI systems may be capable of learning their boundaries and correcting their judgment on their own, but the same is true for humans, and we still impose criminal liability on humans for strict liability offenses. Consequently, if AI entities have the required capabilities for meeting both factual and mental elements of criminal liability for strict liability offenses, and if the rationale for the imposition of criminal liability for these offenses is relevant to both humans and AI systems, then there is no reason to avoid imposing criminal liability in all these cases. But this is not the only type of AI involvement in criminal liability for strict liability offenses.

4.3.2. Human Liability: Perpetration-through-Another

As described earlier,[35] the most common way of handling the instrumental use of individuals for the commission of offenses is the general form of perpetration-through-another. To impose criminal liability for the perpetration of an offense-through-another, it is necessary to prove awareness of the instrumental use. Therefore, perpetration-through-another is applicable only in *mens rea* offenses. In most cases, the person being instrumentally used by the perpetrator is considered an innocent agent, and no criminal liability is imposed on him or her. We have already analyzed perpetration-through-another in a context of *mens rea* offenses.

The person who is instrumentally used, however, may also be considered a semi-innocent agent who is criminally liable for negligence, al-

Table 4.2 · The Legal State Reflected in the Criminal Mental State of the Other Person in Perpetration-through-Another

CRIMINAL MENTAL STATE OF THE OTHER PERSON	LEGAL STATE OF THE OTHER PERSON
mens rea	accomplice
negligence	semi-innocent agent
strict liability	innocent agent
none	innocent agent

though the perpetrator is at the same time criminally liable for a *mens rea* offense. Negligence is the lowest level of mental element required for the person instrumentally used to be considered a semi-innocent agent. In this context, strict liability is too low a level of mental element for consideration as a semi-innocent agent. If the person who is instrumentally used by another is in a mental state of strict liability, that person is to be considered an innocent agent, as if lacking any criminal mental state at all.

Thus, in perpetration of an offense-through-another, the other (instrumentally used) person may be in four possible mental states that reflect matching legal consequences, as described schematically in table 4.2.

When the other person is aware of the delinquent enterprise and still continues to participate although he is under no pressure, he becomes an accomplice to the commission of the offense. Negligence reduces the person's legal state to that of a semi-innocent agent.[36] But whether the mental state of this person is that of strict liability or an absence of a criminal mental state, he must be considered an innocent agent, so that no criminal liability is imposed on him, and the full criminal liability for the relevant offense is imposed on the perpetrator who instrumentally used that person. This is true for both humans and AI entities.

For this legal construct of perpetration-through-another, it is immaterial whether the AI system was instrumentally used, and whether it used its strong AI capabilities to form a strict liability mental state. Thus, the instrumental use of a weak AI system, of a strong AI system that formed strict liability, or of a screwdriver has the same legal consequences. The entity making the instrumental use (human, corporation, or AI system) is criminally liable for the commission of the offense in full, and the instrumentally used entity (human, corporation, or AI system) is considered an innocent agent.

For example, the human user of an unmanned vehicle based on an advanced AI system instrumentally uses the AI system to cross an intersection where the traffic light is red. The system has the capability to be aware of the traffic light's color, but in practice it is not aware of the red light. The system does not take all reasonable measures to examine the data, such as analyzing the color of the traffic light facing in its direction. Crossing an intersection on a red light is a strict liability offense. In this instance, the human is criminally liable for that offense as perpetrator-through-another. Because the AI system was instrumentally used by the perpetrator, it is considered an innocent agent, and therefore criminal liability is not imposed on it.

4.3.3. Joint Human and AI Entity Liability: Probable Consequence Liability

Probable consequence liability addresses the commission of unplanned offenses (different from a planned offense or additional to it). The question in these cases is about the criminal liability of the other parties to the unplanned offense committed by one party. Previously, we discussed probable consequence liability for unplanned *mens rea* and negligence offenses.[37] The relevant question here is regarding the probable consequence liability for unplanned strict liability offenses committed by an AI system. Are the programmers, users, and other related persons criminally liable for unplanned strict liability offense committed by an AI system?

For example, two people are committing a bank robbery. To escape from the arena, they use an unmanned vehicle based on an advanced AI system. During their escape, the vehicle exceeds the legal speed limit, committing a strict liability offense. Analyzing the records of the vehicle shows that it meets the strict liability requirements of the offense. The bank robbers did not program the vehicle to commit the offense, and did not order it to do so. The question concerns the robbers' criminal liability for the strict liability traffic offense committed by the vehicle, in addition to their criminal liability for robbery. If the offenders had ordered the vehicle to drive at that speed and had instrumentally used it for that purpose, this would have been perpetration-through-another, but that is not the case here.

Probable consequence liability may be relevant here for the criminal liability of the bank robbers. The mental condition for probable consequence liability requires the unplanned offense to be "probable" from the point of view of the party that did not commit it. In other words, it is necessary for that party to have been able to foresee and reasonably predict the commission of the offense. Some legal systems prefer to examine actual and subjective foreseeability (the party has actually and subjectively foreseen the occurrence of the unplanned offense), whereas others prefer to evaluate the ability to foresee based on an objective standard of reasonability (the party has not actually foreseen the occurrence of the unplanned offense, but any reasonable person under the same condition would have).

Actual foreseeability parallels subjective *mens rea*, whereas objective foreseeability parallels objective negligence. A lower level of foreseeability is not adequate for probable consequence liability. Thus, the question is about the level of foreseeability required of the robbers. If they actually foresaw the commission of that offense by the vehicle, then criminal liability for the offense is imposed on them in addition to criminal liability for robbery. If the robbers formed objective foreseeability, then criminal liability for the additional offense is imposed only in legal systems in which probable consequence liability can be satisfied through objective foreseeability.

But if the mental state of the bank robbers in relation to the additional offense is strict liability, this is not adequate for the imposition of criminal liability through probable consequence liability. Although the additional offense requires strict liability for imposition of criminal liability, this is true only for the perpetrator of that offense, and not for imposing criminal liability through probable consequence liability. Thus, if neither subjective nor objective foreseeability regarding the commission of the offense can be attributed to the robbers, then probable consequence liability is irrelevant. In this type of cases, no criminal liability is imposed on the offenders, and the criminal liability of the AI system for the strict liability offense does not affect the offenders' liability.

CLOSING THE OPENING EXAMPLE: STRICT LIABLE KILLING ROBOT

We opened this chapter with an example of a killing robot, an unmanned vehicle based on an AI system, subject to strict liability. Returning to the example, we can now analyze it using the insights gained from this chapter. The criminal liability of three main characters must be considered for the homicide in this example: the AI system, the manufacturer (including the programmers), and the user. To examine the criminal liability of the AI system, first we must investigate its technological capabilities. If the AI system has the technological capabilities needed to meet the requirements of strict liability offenses, it forms a *machina sapiens criminalis* in this context.

If such capabilities are present, the criminal liability of the AI system should be examined as if it were human. But if the AI system lacks the technological capabilities needed to meet the requirements of strict liability offenses, it does not form a *machina sapiens criminalis* in this context and has no criminal liability for the offense. In the opening example, the AI system met the factual element requirement of homicide, because its conduct caused a chain of events that ended in the patient's death.[38]

To meet the mental element requirement, we must prove strict liability, which includes proving that the AI system is capable of consolidating awareness or negligence, and is capable of reasonability. To refute the negligence presumption, it is necessary to prove that the AI system has not actually formed awareness of relevant facts or negligence toward them, and has taken all reasonable measures to prevent the commission of the offense. The first condition may be proven based on the characteristics of the system, using the definitions of awareness and negligence in criminal law discussed earlier.[39] If the system possesses such capability, and if such capability has been activated, the system formed awareness or negligence.

The capability for reasonability is not different from the required parallel capability for negligence, previously discussed.[40] The system is not required to act reasonably, only to possess the capability of making reasonable decisions. Strong AI systems, which can calculate the probabilities of the occurrence of events based on current factual data (such as weather forecast computers[41]) and choose options that meet certain criteria, are supposed to have this type of capability. If no additional evidence is brought before the court, this is sufficient to activate the negligence presumption and to impose criminal liability for the strict liability offense.

Refuting the negligence presumption begins with positive proof of not having formed awareness or negligence. This may be proven based on the records of the AI system. If proven, it is necessary to prove having taken all reasonable measures for the prevention of the offense. Because the capability for reasonability has already been proven, the conduct of the system is examined accordingly. The reasonable options calculated by the system and the actions it took based on these calculations must be examined to determine whether they include all the reasonable options in this case. The examination is carried out the same way as in the case of negligence, at times using expert witnesses. If it is proven that no actual awareness or negligence were formed regarding the factual data, and that all reasonable measures have been taken by the system, the negligence presumption is refuted. If not, criminal liability for the strict liability offense is imposed.

The criminal liability of the manufacturer, the programmer, and the user is determined based on their role in causing the death. If they instrumentally used the AI system to kill the patient in the ambulance by designing the system in such a way that it would collide with the ambulance because it misunderstands the factual situation, they are perpetrators-through-another of homicide. This is also the case if the AI system has no advanced AI capabilities to meet the requirements of the offense. In this case, they and the related parties, but not the AI system, are the only ones criminally liable for the homicide. Their exact criminal liability for the homicide, whether as murder or manslaughter, is determined by their meeting the mental element requirements of these offenses.

If the manufacturer, the programmer, and the user foresaw the consequences (the patient's death) or any reasonable person in their condition could have foreseen these consequences, and if the AI system is criminally liable for the homicide, they may be criminally liable for the homicide through probable consequence liability, as long as the offense was unplanned. Generally, if they did not design the system to commit any offense, but the unplanned homicide occurred, then the criminal liability of the system is examined based on its capabilities and on meeting the relevant homicide requirements, and the criminal liability of the human parties is examined based on the standards of probable consequence liability.

Therefore, AI systems can be criminally liable for strict liability offenses, and that does not necessarily reduce the criminal liability of other relevant entities.

5

APPLICABILITY OF GENERAL DEFENSES TO AI CRIMINAL LIABILITY

Can an AI system be insane? An infant? Intoxicated? Caught in situations of self-defense, necessity, or duress? One of the most common uses of AI technology is for guarding, as noted in chapter 1.[1] For example, the South Korean government uses AI robots as soldiers along its border with North Korea, and since 2012 as prison guards.[2] This technology faces particular threats to the software, aimed at weakening its ability to discharge its guarding functions. This technology may also face dilemmas of the type humans encounter in everyday life. If a prison guard apprehends an escaping prisoner and the only way to stop him is by causing injury, is it right for the robot to injure a human for that purpose?

Consider a case in which hackers manage to break into the main computer controlling robots that function as prison guards. The hackers' purpose is to help a friend escape from prison by incapacitating some of the robots. The hackers plant a virus in the system, which immediately affects the robots, preventing them from forming the correct image of the factual reality, and causes them to ignore some of the rules of the system. Some of the prisoners use this situation to attempt an escape. One prisoner captures a human guard and takes his uniform, trying to escape by wearing it. This prisoner is caught by a human guard, however, and a struggle ensues. A robot witnessing the situation misinterprets it as a struggle between two prisoners, and approaches the two men with the intention of intervening to stop the violence. When the human guard orders the robot to keep away, the robot interprets this as a threat to its mission, and grabbing the guard with its hydraulic arm, the robot smashes him against the wall. The disguised prisoner escapes, and the guard dies. The legal question is, who is to be held criminally liable for the guard's death?

The software controlling these robots is considered strong AI technology, so the robots are capable of consolidating awareness and *mens rea* in the context of criminal law. Analysis of the records of these robots reveals that the *mens rea* requirements of homicide are met in full. But this is not the regular course of conduct for these robots, which are programmed not to be violent unless it is absolutely necessary. Accurate analysis of this case requires the application of the general defenses of criminal law to AI technology. This is the question that this chapter seeks to answer: are general defenses applicable to AI criminal liability?

5.1. THE FUNCTION OF GENERAL DEFENSES IN CRIMINAL LAW

General defenses in criminal law are complementary to the mental element requirement; both address the offender's fault in the commission of the offense. The mental element requirement represents the positive aspect of the fault (what should be present in the offender's mind during the commission of the offense), and the general defenses are the negative aspect of the fault (what should be missing from the offender's mind during the commission of the offense).[3] For example, awareness is part of the mental element requirement (*mens rea*), and insanity is a general defense. Thus, in *mens rea* offenses, the offender must be aware and must *not* be insane.

Therefore, the fault requirement in criminal law consists of the mental element requirement and the general defenses. The general defenses were developed in the ancient world to prevent injustice in certain types of cases. For example, someone who killed another out of self-defense was not criminally liable for the homicide, because the person lacked the required fault to cause death. An authentic factual mistake by the offender regarding the intentional commission of an offense was considered as negating the required fault for imposing criminal liability.[4] In the modern era, the general defenses became wider and more conclusive, but the common factor of all general defenses remained the same.

All general defenses in criminal law are part of the negative aspect of the fault requirement, as they are intended to negate the offender's fault. The deep abstract question behind the general defenses is whether the commission of the offense was not imposed on the offender in some way. For example, when a person really acts in self-defense, the offense is considered to be imposed on her. To save her own life, which is considered a legitimate purpose, the offender had no choice but to act in self-defense. Naturally, she could have chosen to give up her life, but that is not considered to be a legitimate requirement because it goes against the natural instinct of every living creature.

General defenses can be divided into two main types: exemptions and justifications.[5] Exemptions are general defenses related to the personal characteristics of the offender (*in personam*), and justifications are related to the characteristics of the factual event (*in rem*). The personal characteristics of the offender may negate the fault for the commission of the offense, regardless of the factual characteristics of the event or the exact identity of the offense. In exemptions, the personal characteristics of the offender are sufficient to prevent imposition of criminal liability for any offense.

For example, a child under the age of legal maturity is not criminally liable for any offense factually committed by him. The same is true for individuals who committed the offense at a time when they were considered to be insane. The exact nature of the offense committed by the individual is immaterial for the imposition of criminal liability. It may be relevant for subsequent steps toward the treatment or rehabilitation of the offender, but not for the imposition of criminal liability.

Justifications are impersonal general defenses, and as such they do not depend on the identity of the offender, but only on the factual event that took place. The personal characteristics of the individual are immaterial for justifications. For example, one person is attacked by another person, and her life is in danger. She pushes the attacker away, which is considered assault unless it is done with consent. In this case, the person doing the pushing claims self-defense regardless of her identity, the identity of the attacker, or any other of their personal attributes, because self-defense has to do with only the factual event itself.

Because justifications are impersonal, they also have a prospective value. Not only is the individual not criminally liable if she acted with justification, but this is the way in which she should have been acting. Justifications define not only types of general defense, but also proper behavior.[6] Therefore, individuals should defend themselves in situations that require self-defense, even if this may appear as an offense. This is not true for exemptions. A child below the maturity age is not supposed to commit offenses even if no criminal liability is imposed on him, and neither are insane individuals.

The prospective behavioral value of justifications expresses the social values of society. If we accept self-defense, for example, as legitimate justification, this means that we prefer people to protect themselves when the authorities are unable to protect them. We prefer to reduce the monopoly of the state on power by legitimizing self-assistance rather than leave people vulnerable and helpless. We do not coerce individuals to act in their self-defense, but if they do, then we do not consider them criminally liable for the offense committed through self-defense.

Both exemptions and justifications are general defenses. The term *general defenses* relates to defenses that may be applied to any offense, and not to a certain group of offenses. For example, infancy can be applied to any offense as long as it was committed by an infant. By contrast, some specific defenses can be applied only to specific offenses or types of offenses. For example, in some countries in the case of statutory rape, a defense is applicable if the age gap between the defendant and the victim is less than three years. This defense is unique to statutory rape, and it is irrelevant for any other offense. Exemptions and justifications are classified as general defenses.

As defense arguments, general defenses are argued positively by the defense. If the defense chooses not to raise these arguments, they are not discussed in court, even if participants at the trial understand that such an argument may be relevant. It is not sufficient to state the general defense: its elements must be proven by the defendant. In some legal systems, the general defense is proven by raising a reasonable doubt about the presence of the elements of the defense. In other legal systems, the general defense must be proven by a preponderance of the evidence. Thus, the prosecution has the opportunity to refute the general defense.

General defenses of the exemption type include infancy, loss of self-control, insanity, intoxication, factual mistakes, legal mistakes, and substantive immunity. General defenses of the justification type include self-defense (including defense of dwelling), necessity, duress, superior orders, and *de minimis* defense. All these general defenses may invalidate the offender's fault. The question is whether these general defenses are applicable to AI technology in the context of criminal law. We answer this question now, discussing exemptions and justifications separately.

5.2. EXEMPTIONS

Exemptions are general defenses related to the personal characteristics of the offender (*in personam*), as noted previously.[7] The applicability of exemptions to the criminal liability of AI entities raises the question of the capability of these entities to form the personal characteristics required for general defenses. For example, the question of whether an AI system could be insane is rephrased to ask whether it has the mental capability of forming the elements of insanity in criminal law. This question, *mutatis mutandis*, is relevant to all exemptions, as we will discuss.

5.2.1. Infancy

Could an AI system be considered an infant for the purposes of criminal liability? Since ancient times, children under a certain biological age were not considered criminally liable (*doli incapax*). The difference between the various legal systems was in the exact age of maturity. For example, Roman law set the age of maturity at seven years.[8] This defense is determined by legislation[9] and case law.[10] It is clear that the cutoff age is biological and not mental, primarily for evidentiary reasons, because biological age is much easier to prove.[11] It was presumed, however, that the biological age matches the mental age.

If the child's biological age exceeds the lower threshold but is under the age of full maturity (for example, fourteen in the United States), the mental age of the child is examined based on evidence (for example, expert testimony).[12] The conclusive examination is whether the child understands his own behavior and its wrong character,[13] and whether he understands that he may be criminally liable for the offense, as if he were mature. But there may still be some procedural changes in the criminal process compared to the standard process (for example, juvenile court, the presence of parents, or more lenient punishments).

The rationale behind this general defense is that children under a certain age, whether biological or mental, are presumed to be incapable of forming the fault required for criminal liability.[14] The child is not incapable of mentally containing the fault and understanding the full social and individual meanings of criminal liability, so that imposition of criminal liability would be irrelevant, unnecessary, and vicious. Consequently, children are not criminally liable but rather are educated, rehabilitated, and treated.[15] The question, in our case, is whether this rationale is relevant only to humans but to other legal entities as well.

The general defense of infancy is not considered applicable to corporations. There are no "child corporations," so the moment a corporation is registered it exists, and criminal liability may be imposed on it legally; therefore, the rationale of this general defense is irrelevant for corporations. A child lacks the mental capability to form the required fault because human consciousness is underdeveloped at this age. As the child becomes older, his or her mental capacity develops gradually until it reaches full understanding of right and wrong. At this point, criminal liability becomes relevant.

The mental capacity of corporations does not depend on their chronological "age," that is, the date of registration, and it is considered to be constant. Moreover, the mental capacity of a corporation derives from its human officers, who are mature entities. Consequently, there is no legitimate justification for the general defense of infancy to be applied to corporations. The question of interest to us is whether AI systems resemble humans or corporations in this context.

The answer differs for different types of AI systems, and we must distinguish between fixed AI systems and dynamically developing ones. Fixed AI systems begin their activity with the capacities they are going to have throughout their life cycle. These systems do not experience any change in their capacities with the passage of time. Consequently, their capacity to form mental requirements (for example, awareness, intent, or negligence) must be examined at every point where an offense has been committed, and the general defense of infancy does not apply.

But the starting and end points of dynamically developing AI systems are different. Their capacities, including mental ones, develop over time through machine learning or other techniques. If the system began its activity without the mental capacities required for criminal liability, and at some point it developed such capacities, then the time between the starting point and the point when it possesses such capacities parallels that of childhood. During this period, the system does not have the capabilities required for the imposition of criminal liability.

But if the mental capacity of the AI system is already developed at the time the offense is committed, the question arises whether the infancy defense is still relevant because if the system possesses the required capabilities, it is criminally liable regardless of the point in time when these capabilities became available. The answer to this question is similar to the rationale for the general defense for human offenders.[16] We can examine the mental capacity of each child at any age and determine individually whether the child possesses the required capabilities, but this would be very inefficient. Any preschooler who may have hit another one in day care is immediately identified as not having the mental capacities required for criminal liability.

The other way is to establish a legal presumption whereby children under a certain age are not criminally liable. This is the situation with AI systems. Massive production of strong AI systems (for example, dozens of prison guards or military robots), having the exact same capabilities and using the same learning techniques, makes it unnecessary to evaluate the mental capacities of each individual robot. If we know empirically that a given system possesses all required mental capabilities after 264 hours of activity, and if such a robot commits an offense before having operated for 264 hours, the robot is presumed to be in its childhood and no criminal liability is imposed.

If the prosecution insists that the individual system possesses the required mental capacities already in its "infancy," or if the defense insists that the system does not possess the required mental capacities despite the fact that it has passed its infancy, the actual mental capacities of the systems can be examined and evaluated specifically. This would not be substantively different from the natural gaps between the biological and mental ages of humans. When the argument is made that a seventeen-year-old human offender is mentally underdeveloped, the court examines the offender's mental capacities and decides whether that person has the required capacity to become criminally liable.

This rationale is relevant for both humans and dynamically developing AI systems, but not for corporations. Therefore, it seems that the general defense of infancy can be relevant for this type of AI systems under the right circumstances.

5.2.2. Loss of Self-Control

Could an AI system experience loss of self-control in the context of criminal liability? Loss of self-control is a general defense that has to do with one's inability to control one's bodily movements. When the reason for this inability is mental disease, it is considered insanity,[17] and when it is the effect of intoxicating substances, it is considered intoxication.[18] But the general defense of loss of self-control is more general than insanity and intoxication because it does not require a specific type of reason for the loss of self-control. Whenever the offender's bodily movements are not under his or her full control, this defense may be used if its conditions are met.[19]

The general rationale of this defense is that uncontrollable bodily movement does not reflect the offender's will, and therefore it should not be the basis for the imposition of criminal liability. Thus, the uncontrollable reflexes of an individual may be an expression of loss of self-control.[20] For example, when a physician, during a routine medical checkup, taps a patient's knee, she triggers a reflex that causes the patient's leg to move forward. The leg may end up kicking the physician, which ordinarily may have been considered assault, but because in this instance it is the result of a reflex, the general defense of loss of self-control is applicable, and no criminal liability is imposed.

But for the general defense to be applicable, the loss of self-control must be total. In this example, if the patient tapped his own knee on purpose for the reflex to be triggered, and as a result his leg kicked the physician, the general defense is not applicable because the assault reflects the offender's will. Consequently, two accumulative conditions must be met for the loss of self-control defense to be applicable: (1) inability to control the self-behavior, and (2) inability to control the conditions under which that behavior occurred.

In our example, the patient did not control his self-behavior (the reflex), but he controlled the conditions for its occurrence (tapping the knee); therefore, the general defense of loss of self-control is not applicable to him.

Many types of situations have been recognized as loss of self-control, including automatism (acting without aware central control over the body),[21] convulsions, post-epileptic states,[22] post-stroke states,[23] organic brain diseases, diseases of the central nervous system, hypoglycemia, hyperglycemia,[24] somnambulism (sleepwalking),[25] extreme sleep deprivation,[26] side effects of bodily[27] or mental traumas,[28] blackout situations,[29] side effects of amnesia[30] and brainwashing,[31] and many more.[32] The cause for the loss of self-control is immaterial for meeting the first condition. As long as the offender is indeed incapable of controlling his or her behavior, the first condition is fulfilled.

The second condition refers to the cause for being in the first condition. If the cause was controlled by the offender, he is not considered to have lost his self-control. Controlling the conditions for losing the self-control is controlling the behavior. Therefore, when the offender controls the conditions for moving in and out of control, he cannot be considered to have lost his self-control. In Europe, the second condition is based on the doctrine of *actio libera in causa*, which dictates that if the one controls one's entry into a situation that is out of control, the general defense of loss of self-control is rejected.[33]

Accordingly, AI systems can experience loss of self-control in the context of criminal law due to external and internal causes. For example, a human pushes an AI robot onto another human. The robot being pushed has no control over that movement. This is an example of an external cause for loss of self-control. If the pushed robot makes nonconsensual physical contact with the other person, this may be considered an assault. The mental element required for assault is awareness, so if the robot is aware of that physical contact, both factual and mental elements requirements of the assault are met.

If the AI robot were human, it would have probably claimed the general defense of loss of self-control. Thus, although both mental and factual elements of the assault are fulfilled, no criminal liability is imposed because commission of the offense was involuntary, or due to loss of self-control. This general defense would prevent the imposition of criminal liability on human offenders, and it should also prevent the imposition of criminal liability on robots. If the robot has no AI capabilities for consolidating awareness, there is no need for this defense because the robot would be functioning as a mere tool. The AI robot, however, is aware of the assault.

The ability of the AI robot to meet the mental element requirement of the offense makes it necessary to apply the general defense, which functions identically with humans and robots. An example of internal cause for loss of self-control is the case of an internal malfunction or technical failure that causes uncontrolled movements by the robot. The robot may be aware of the malfunction and still not be able to control it or correct it. This is also the case for the general defense of loss of self-control. Whether the cause for the loss of self-control is external or internal is relevant to the applicability of this general defense.

If the robot controlled these causes, however, the defense is not applicable. For example, if the robot physically caused a person to push it (the robot) onto another person (external cause), or if the robot caused the malfunction knowing the probable consequences of this for its mechanism (internal cause), then the second condition of the defense is not met and the defense is not applicable. The situation is the same for humans. It appears, therefore, that the general defense of loss of self-control is applicable to AI systems.

5.2.3. Insanity

Can an AI system be considered insane for the purposes of criminal liability? Insanity has been known to humanity since the fourth millennium BC,[34] but at that time it was considered to be a punishment for religious sins,[35] and therefore there was no need to look for cures.[36] Only since the middle of the eighteenth century has insanity been explored as a mental disease, and its legal aspects considered.[37] In the nineteenth century, the terms *insanity* and *moral insanity* described situations in which the individual lost his or her moral orientation, or had a defective moral compass, but was aware of common moral values.[38]

Insanity was diagnosed as such only in case of significant deviations from common behavior, especially sexual behavior.[39] Since the end of the nineteenth century, it has been understood that insanity is mental malfunctioning, and that at times it is not expressed in deviations from common behavior. This approach lies at the foundation of the understandings of insanity in criminal law and criminology.[40] Mental diseases and defects were categorized according to their symptoms and respective medical treatments, and their effect on criminal liability was explored and recorded. But the different needs of psychiatry and criminal law produced different definitions for insanity.

For example, the early English legal definition of an insane person ("idiot") was not being able to count to twenty. In psychiatry, such a person is not considered insane.[41] Criminal law needed a clear and conclusive definition of insanity, whereas psychiatry had no such needs. In most modern legal systems, today's legal definition of insanity is inspired by two nineteenth-century English tests. One is the *M'Naughten* rules of 1843,[42] and the second is the irresistible impulse test of 1840.[43] The combination of these two tests ensures that the general defense of insanity complies with the structure of *mens rea*, as discussed previously.[44]

The legal definition of insanity has both cognitive and volitive aspects. The cognitive aspect of insanity addresses the ability to understand the criminality of the conduct, whereas the volitive aspect addresses the ability to control the will. Thus, if a mental disease or defect causes cognitive malfunction (difficulty to understand the factual reality and the criminality of the conduct) *or* volitive malfunction (irresistible impulse), the condition is considered insanity from the legal point of view.[45] This is the conclusive common test for insanity.[46] It fits the structure of *mens rea*, which also contains both cognitive and volitive aspects, and it is complementary to the *mens rea* requirement.[47]

This definition of insanity is functional and not categorical. For an individual to be considered insane, it is not necessary to have a mental illness that appears on a certain list of mental diseases. Any mental defect, of any type, can be the basis for insanity as long as it causes cognitive *or* volitive malfunctions. The malfunctions are examined functionally and not necessarily medically, and they need not appear on a list of mental diseases.[48] Therefore, a person may be considered insane by criminal law and perfectly sane from the point of view of psychiatry, as in the case of a cognitive malfunction that is not categorized as a mental illness. The opposite situation, whereby a person is sane for the purposes of criminal law but insane psychiatrically, is also possible, as in the case of a mental disease that does not cause any cognitive or volitive malfunction.

The insane person is presumed to be incapable of forming the fault required for criminal liability. The question is whether the general defense of insanity is applicable to AI systems. The general defense of insanity requires a mental, or inner, defect that causes cognitive or volitive malfunction. There is no need to demonstrate any specific type of mental disease, and any mental defect is satisfactory. The question is, how can we know about the existence of that "mental defect"? Because the mental defect is examined functionally and not based on certain categories, the symptoms of that mental defect are critical for its identification.

The inner defect causes cognitive or volitive malfunction, whether it is classified as a mental disorder, or a chemical imbalance in the brain, or an electric imbalance in the brain, and so on. The inner cause is examined based on its functional effect on the human mind. This is the legal situation with humans, and the same holds true for AI systems. The more complicated and advanced the AI system, the higher the probability of inner defects mostly in the software, but also in the hardware. Some inner defects do not cause the AI system to malfunction, but others do. If the inner defect causes a cognitive or volitive malfunction of the AI system, this matches the definition of insanity in criminal law.

Because strong AI systems are capable of forming all *mens rea* components, as we have discussed,[49] and given that these components consist of cognitive and volitive components owing to the *mens rea* structure, it is quite likely that some inner defects can cause these capabilities to malfunction. When an inner defect causes such a malfunction, this matches the definition of insanity in criminal law. Partial insanity is applicable when the cognitive or volitive malfunctions are not complete. Temporary insanity is applicable when these malfunctions affect the offender (human or AI system) for a certain period of time.[50]

It is possible to argue that this is not the typical character of the insane person because it does not match the concept of insanity reflected in psychiatry, culture, folklore, literature, and even the movies. Nevertheless, it is insanity from the perspective of criminal law. First, the criminal law definition of insanity differs from its definitions in psychiatry, culture, and so on, and it is the definition that is used for humans. There is no reason that a different definition should be used for AI systems. Second, criminal law does not require a mental disease for human insanity, so why should we require it for AI systems?

The definition of insanity used by criminal law may seem too technical, but if the offender meets its requirements, it is applied. If both human and AI offenders are capable of meeting the insanity requirement in criminal law, there is no legitimate reason for making the general defense of insanity applicable for one but not the other. Consequently, it appears that the general defense of insanity is applicable to AI systems.

5.2.4. Intoxication

Can an AI system be considered intoxicated for the purposes of criminal liability? The effects of intoxicating substances have been known to humanity since prehistory. In the early law of ancient times, the term *intoxication* referred to drunkenness as a result of ingesting alcohol. Later, when the intoxicating effects of other materials became known, the term was expanded.[51] Until the beginning of the nineteenth century, intoxication was not accepted as a general defense. The Archbishop of Canterbury wrote in the seventh century that imposing criminal liability on a drunk person who committed homicide was justified for two reasons: first, the drunkenness itself, and second, the homicide of a Christian person.[52]

Drunkenness was conceptualized as a religious and moral sin, and therefore it was considered immoral to let offenders be exempt from criminal liability for being drunk.[53] Only in the nineteenth century did the courts undertake a serious legal discussion of intoxication, made possible by legal and scientific developments that have produced the understanding that an intoxicated person is not necessarily mentally competent for purposes of criminal liability (*non compos mentis*).

From the beginning of the legal evaluation of intoxication in the nineteenth century, the courts distinguished between cases of voluntary and involuntary intoxication.[54] Voluntary intoxication was considered to be the offender's fault, so it could not be the basis for exemption from criminal liability. Nevertheless, voluntary intoxication could be considered as a relevant circumstance for the imposition of a more lenient punishment.[55] Moreover, voluntary intoxication could be used to refute premeditation in first-degree murder cases.[56] Courts have distinguished between cases based on the reasons for becoming intoxicated. Voluntary intoxication grounded in the will to commit an offense was considered different from voluntary intoxication undertaken for no criminal reason.[57]

By contrast, involuntary intoxication was recognized and accepted as an exemption from criminal liability.[58] Involuntary intoxication is a situation imposed on the individual, so it is not just and fair to impose criminal liability in such situations. Thus, the general defense of intoxication has two main functions. When intoxication is involuntary, the general defense prevents imposition of criminal liability. When it is voluntary but not intended for the commission of an offense, intoxication is a consideration for the imposition of a more lenient punishment.

The modern understanding of intoxication includes any mental effect caused by an external substance (for example, chemicals). The required mental effect matches the structure of *mens rea*, discussed earlier. Consequently, the effect may be cognitive or volitive.[59] The intoxicating effect may relate to the offender's perception, understanding of factual reality, or awareness (cognitive effect); or it may affect the offender's will, including the creation of an irresistible impulse (volitive effect). Intoxication is caused by an external substance, but there is no closed list of substances that may be illegal (heroin, cocaine, and so on) or perfectly legal (alcohol, sugar, pure water, and so on).

The effect of the external substance on the individual is subjective, and various people may be affected differently by the same substances in the same quantities. Sugar may produce hyperglycemia in one person, but another person is barely affected by it. Pure water may result in an imbalance of the electrolytes in one person, but another is barely affected by it. Cases of addiction raised the question whether the *absence* of the external substance may also be considered a cause for intoxication. For example, when a person addicted to narcotics is in withdrawal, he or she may experience cognitive and volitive malfunctions as a result of the drug's absence.

Consequently, the cognitive and volitive effects of narcotics addiction are considered intoxication by criminal law.[60] To determine the question of voluntary or involuntary intoxication (in case the addicted person wanted to begin the process of weaning), it is the cause for the addiction, not for the weaning, that is examined to be voluntary or involuntary.[61] Thus, intoxication is examined through a functional evaluation of its cognitive and volitive effects on the individual, regardless of the exact identity of the external substance that is the cause of these effects. The question is whether the general defense of intoxication is applicable to AI systems.

As already noted, the general defense of intoxication requires an external substance (for example, the presence or absence of a certain chemical) that has cognitive or volitive effects on the inner process of consciousness. For example, the manufacturer of AI robots wanted to reduce production costs, and used inexpensive materials. After a few months of operation, a process of corrosion began in some of the initial components of the robots, and as a result, transfer of information was deficient and affected the awareness process. Technically, this is similar to the effect of alcohol on human neurons.

In another example, a military AI robot was activated in a civilian area after a real or simulated chemical attack. As a result of exposure to the poisonous gas, parts of the robot's hardware were damaged, and the robot began to malfunction, became unable to properly identify people, and started to attack innocent civilians. Subsequent analysis of the robot's records showed that exposure to the gas was the only reason for the attacks. If the robot had been human, the court would have accepted the general defense of intoxication and acquitted the defendant. Should not the same procedure also apply to AI systems?

If there is no difference between the effects of external substances on humans and AI systems from a functional point of view, then there is no justification for applying the general defense of intoxication to one and not the other. Strong AI systems can possess both cognitive and volitive inner processes, as noted previously.[62] These processes can be affected by various factors, and when they are affected by external substances, as in the examples just presented, the requirements of intoxication as a general defense are met. Thus, if exposure to certain substances affects the cognitive and volitive processes of an AI system in a way that causes it to commit an offense, there is no reason that intoxication as a general defense should not be applicable.

It may be true that AI systems cannot become drunk on alcohol or have delusions following the ingestion of drugs, but these effects are not the only possible ones related to intoxication. If a human soldier attacks his friends as a result of exposure to a chemical attack, his argument for intoxication is accepted. If this exposure has the same substantive and functional effect on both humans and AI systems, there is no legitimate reason for making the general defense of intoxication applicable to one type of offender but not another. Consequently, it appears that the general defense of intoxication can be applicable to AI systems.

5.2.5. Factual Mistake

We are such stuff
As dreams are made on, and our little life
Is rounded with a sleep.
   —William Shakespeare, *The Tempest*, 1611

Can an AI system be considered factually mistaken for the purposes of criminal liability? The general defense of factual mistake provides a revised perspective of the cognitive aspect of consciousness. In our discussion of awareness, the assumption was that there is a factual reality of which the individual may or may not be aware.[63] This reality was considered constant, objective, and external to the individual. But the only way an individual can know about the factual reality is through the process of awareness, that is, perception by senses and understanding of the factual data.[64]

Only when our brain tells us that this is the factual reality do we believe it; we have no other way to evaluate reality's existence. But sights, sounds, smells, or physical pressure may be simulated, and the human brain may be stimulated to feel them, although they may not necessarily exist. The ultimate example is dreaming. Often, we dream without knowing that we are dreaming, and we cannot distinguish between dream and factual reality while we are dreaming. In our dreams we see sights, hear sounds, smell, feel, talk, or run as if we were facing factual reality.

What makes the dream different from factual reality in the eyes of the dreamer? For most people, the dream is a dream simply because the dreamer eventually exits it by waking up. But what would happen if, on the contrary, we were to wake up in a dream? This option is not accepted by most people because our intuition tells us that if we open our eyes, if we are in bed, and if it is morning, the dream portion of this day has ended and the waking portion begins. But when we are in the midst of the dream, we have no reliable way of distinguishing dreams from what we call "factual reality" (within quotation marks, because we do not have a reliable way of verifying it).

We "see" things with our brain, not with our eyes. Our eyes are merely light sensors. The light is converted into electrical currents and chemical processes that carry messages between neurons and form the "sight" by stimulating the brain in the right spots. But it is possible to imitate this stimulation. Electrodes can be attached to the brain to stimulate it in the appropriate regions and make it "see," "hear," or "feel." The stimulated brain becomes aware of these images, sounds, and so on, and people would consider them factual reality.

If we begin to doubt our awareness of the factual reality, we must also consider the problem of perspective. Even if we assume that we experience the factual reality (without quotation marks), we can experience it only through our subjective perspective, which is not necessarily the only perspective of that reality. For example, if we cut out a triangular shape from a piece of cardboard, we may perceive a triangle if we view it facing its flat surface; but if we rotate it by 90 degrees, we perceive it as a straight line. The issue of perspective can be crucial if we associate different interpretations with the different perspectives.

For example, two people see a man holding a long knife and hear him telling a woman that he is about to kill her. One person understand this as a serious threat to the woman's life, and calls police or attempts to intervene and save the woman; the other person understands this as part of a show (for example, a street performance) that requires no intervention. The deep question in criminal law in this context is, what should be the factual basis of criminality: the factual reality as it actually occurred, or what the offender believed to be the factual reality, although it may not have actually taken place?

For example, a man charged with rape admits that he and the complainant had sexual intercourse, but proves that he had really believed that it was carried out consensually; and the prosecution proves that the complainant did not consent. The court believes both, and both are telling the truth. Should the court acquit or convict the defendant? Since the seventeenth century, modern criminal law has preferred the subjective perspective of the individual (the defendant) about the factual reality to the factual reality itself.[65] The general concept is that the individual cannot be criminally liable except for "facts" that he or she believed to be knowing, regardless of whether they actually occurred in factual reality.[66]

In most legal systems, the limitations imposed on this concept are evidentiary, so that the defendant's argument would be considered true and authentic, if proven by the defendant. When the argument is considered true and authentic, this becomes the basis for the imposition of criminal liability. In our example, the defendant is acquitted because he truly believed the intercourse was consensual. If the defendant's perspective negates the mental element requirement, the factual mistake works as a general defense that leads to the defendant's acquittal.[67]

The general defense of factual mistake is applicable not only in *mens rea* offenses, but also in negligence and strict liability offenses. The difference between them is in the type of mistake required. In *mens rea* offenses, any authentic mistake negates awareness of the factual reality and is considered adequate for that general defense. In negligence offenses, the mistake is also required to be reasonable for the defendant to be considered to have acted reasonably.[68] In strict liability offenses, the mistake must be inevitable despite the defendant having taken all reasonable measures to prevent it.[69] The question is whether the general defense of factual mistake is applicable to AI systems.

Both humans and AI systems can experience difficulties, errors, and malfunctions in the process of awareness of the factual reality. These difficulties may occur both in the process of absorbing the factual data by the senses and in the process of creating the relevant general image about the data. In most cases, such malfunctions result in the formation of an inner factual image that differs from the factual reality as the court understands it. This is a factual mistake concerning the factual reality. Factual mistakes are part of our everyday life, and they are a common basis for our behavior.

In some cases, factual mistakes by both humans and AI systems can lead to the commission of an offense. This means that according to factual reality the action is considered an offense, but not so according to the subjective inner factual image of the individual, which happens to involve a factual mistake. For example, a human soldier mistakenly identifies his friend as an enemy and shoots him. For reasons unknown, the soldier who was shot wore the enemy's uniform, spoke the enemy's language, and looked as if he intended to attack the soldier who shot him. Although he was asked to identify himself, he did not comply with the request. In this case, the mistake is authentic, reasonable, and inevitable.

If the human soldier claims a factual mistake, he will probably be acquitted (if indicted at all). If the soldier is not a human but a strong AI robot, should not the criminal law treat the robot soldier the same way that it treats the human soldier? The errors committed by the human and robot soldiers are substantively and functionally identical. The factual mistakes of both humans and AI entities have the same substantive and functional effects on cognition and on the perception of factual reality. As a result, there is no reason for preventing the use of the factual mistake as general defense for AI entities exactly as it is applied to humans.

Computers do make mistakes. The probability of a mistake in mathematical calculations by a computer may be low, but if the computer absorbs mistaken factual data, the results of the calculations may be wrong. Calculations regarding required, possible, and impossible courses of action are affected by the data being absorbed, exactly as in the case of humans.[70] If factual mistakes have identical substantive and functional effects on humans and AI systems, there is no legitimate reason for making the general defense of factual mistake applicable to one type of offender and not to the other. Consequently, it appears that the general defense of factual mistake can be applied to AI systems.

5.2.6. Legal Mistake

Could an AI system be considered legally mistaken for the purpose of criminal liability? A legal mistake is a situation in which the offender either misinterprets the law or is ignorant of it. The general idea behind this defense is that a person who does not know about a certain prohibition, and consequently commits an offense, does not consolidate the required fault for the imposition of criminal liability. Nevertheless, the mental element of the offenses does not include knowing about the prohibition.

For example, the offense of rape requires a mental element of *mens rea*, which includes awareness of the commission of sexual intercourse with a woman and of the absence of consent. That offense does not require that the rapist know that rape is prohibited as a criminal offense. The mental element requirement of rape is satisfied by awareness of the factual element components, regardless of the rapist's knowledge that rape is prohibited. This is the legal situation for most offenses owing to prospective considerations of everyday life in society.[71]

If offenders were required to know about the prohibition as a condition for the imposition of criminal liability, they would be encouraged not to learn the law, and as long as they were ignorant of the law, they would enjoy immunity from criminal liability. If no such condition is required, the public is encouraged to inquire about the law, to know it, and to obey it. These prospective considerations do not fully match the fault requirement in criminal law. As a result, criminal law seeks a balance between justice, the fault requirement, and prospective considerations concerned with everyday life in society.

Initially, considerations were entirely prospective. Roman law stated that ignorance of the law does not excuse the commission of offenses (*ignorantia juris non excusat*),[72] and until the nineteenth century, this was the general approach followed by most legal systems.[73] In the nineteenth century, when the culpability requirement in criminal law underwent a dramatic development, it became necessary to establish a balance, and the legal mistake was required to be made in good faith (*bona fide*)[74] and to reflect the highest state of mental element, that is, strict liability. According to this standard, the legal mistake is an inevitable one, despite the fact that all reasonable measures have been taken to prevent it.[75]

This high standard for legal mistakes is required for *all* types of offenses, regardless of their mental element requirement. Therefore, the general standard for legal mistakes is higher than that for factual mistakes. In this context, the main question in courts is whether the offender has indeed taken all reasonable measures to prevent the legal mistake, including reasonable reliance on statutes, judicial decisions,[76] official interpretations of the law (including pre-rulings),[77] and the advice of private counsel.[78] The question for us is whether the general defense of legal mistake is applicable to AI systems.

Technically, if the relevant entity, whether a human or an AI system, is capable of fulfilling the mental element requirement of strict liability offenses, the entity also is capable of claiming legal mistake as a general defense. Because strong AI systems have the ability to meet the mental element requirement of strict liability offenses, as discussed earlier,[79] they also have the capabilities needed to make the general defense of legal mistake applicable to them. The absence of the legal knowledge of a given issue can be proven by examining the records attesting to the knowledge of the AI entity, thereby meeting the good faith requirement as well.

The fundamental meaning of the applicability of the legal mistake defense to AI systems is that the system was not restricted by any formal legal restriction, and it acted accordingly. If the AI system contains a software mechanism that searches for such restrictions, and although the mechanism has been activated no such legal restriction has been found, then the general defense applies. Note, however, that the system's exemption from criminal liability does not function as an exemption from criminal liability for the programmers or users of the system. If these persons could have restricted the activities of the system to strictly legal ones but failed to do so, they may be criminally liable for the offense based on perpetration-through-another or on probable consequence, as we have discussed.[80]

For example, an AI system absorbs factual data about certain people, and it is required to analyze their personalities based on the data and publish the results within certain guidelines. In one instance, the publication is considered criminal libel. If the records of the system show that it has not been constrained by any restriction regarding libelous publications and it either did not have the mechanism to search for such restrictions or it did have the mechanism but found no such restrictions, then the system is not criminally liable for libel. The manufacturer, programmers, and users may be criminally liable for the libel, however, as perpetrators-through-another or through probable consequence liability.

An AI system may have broad knowledge about many types of issues, but it may not contain legal knowledge on every legal matter. The system may be searching for legal restrictions if it is designed to do that, but may not necessarily find any; this then becomes a case for the general defense of legal mistake. If legal mistakes have the same substantive and functional effects on both humans and AI systems, there is no legitimate reason to make the general defense of legal mistake applicable to one type of offender but not to the other. Consequently, it appears that the general defense of legal mistake is applicable to AI systems.

5.2.7. Substantive Immunity

Could an AI system have substantive immunity in the context of criminal law? Certain types of people enjoy substantive immunity from criminal liability by virtue of their office (*ex officio*). The immunity is granted to these people *ex ante* so that they are not troubled with issues of criminal law related to their office. Society grants these immunities because the offices held by such people are regarded to be much more important than the probable criminal offenses they may commit in the course of discharging their duties. This immunity is not absolute, and it relates to offenses committed by these people as part of discharging their official duty and for its sake.

For example, a firefighter on a mission to save a woman's life is hoisted by a crane and breaks the window to enter her apartment. This act apparently meets the factual and mental element requirements of several offenses (for example, break-in, entering, and property damage), but to enable the firefighter to discharge his lifesaving duty, he is granted substantive immunity from criminal liability for these offenses. If, however, the firefighter had entered the apartment and raped the woman, the immunity would have been immaterial because rape is not part of his duties.

This type of immunity is substantive as it nullifies the criminal liability of the person to whom it is extended, but it does not prevent indictment or other criminal proceedings. The law explicitly lists the types of people to whom substantive immunity is extended (firefighters, police officers, soldiers, and so on). The general defense applies to these people only if they have committed the offense as part of their official duty, in order to discharge that duty, and in good faith (*bona fide*), that is, not exploiting the immunity to deliberately commit other criminal offenses. The question is whether the general defense of substantive immunity is applicable to AI systems.

The firefighter in our example was human, and if indicted for causing property damage would claim substantive immunity. The court would most likely accept the claim and acquit him immediately. Let us assume, however, that the fire posed too heavy a risk for a human firefighter, and an AI robot was dispatched to save the woman's life. The robot is equipped with a strong AI system and capable of making the same decision that a human firefighter would make under the circumstances. Is there any reason such a robot should not be granted similar immunity to that enjoyed by the human firefighter?

If all the conditions for granting this general defense are met, there is no reason for using different standards toward humans and AI systems. AI robots are already being used for official duties (for example, as guards),[81] and inevitably at times they must commit offenses in the course of discharging their duties. For example, prison guards may physically assault escaping prisoners to prevent them from escaping. If these situations have the same substantive and functional effects on both humans and AI systems, then there is no legitimate reason for make the general defense of substantive immunity applicable to one type of offender but not to the other. Consequently, it appears that the general defense of substantive immunity is applicable to AI systems.

5.3. JUSTIFICATIONS

Justifications are general defenses having to do with the characteristics of the factual event (*in rem*), as previously noted.[82] The applicability of justifications to the criminal liability of AI systems raises the question of the capability of these systems to be involved in such situations as self-defense, necessity, or duress. Is it legitimate to allow a robot to defend itself from attack? And if the attack is conducted by humans, is it legitimate to allow a robot to attack humans to ensure the robot's safety?

In general, because justifications are *in rem* general defenses, the personal characteristics of the individual (human or AI system) should be considered immaterial. But these general defenses were designed for humans, taking into account human weaknesses and allowing for them. For example, self-defense was designed to protect the human instinct to protect one's life. Is this instinct relevant to AI systems, which are machines? If not, why would self-defense apply to machines? We will now explore the applicability of justifications as general defenses for AI systems.

5.3.1. Self-Defense

Self-defense is one of the most ancient defenses in human culture. Its basic function is to mitigate the *society's absolute monopoly on power* according to which only society (the state) has the authority to use force against individuals.[83] When one individual has a dispute with another, he or she is not allowed to use force, but must apply to the state (through the courts, police, and so on) for the state to solve the problem and use force if necessary. According to this concept, no power is left in the hands of individuals.

For this concept to be effective, however, state representatives must be present at all times in all places. If one individual is attacked by another in a dark alley, she cannot retaliate, but must wait for the state representatives, who may be unavailable at that time. To enable people to protect themselves from attackers in situations like this, society must partially retreat from this concept. One such retreat is the acceptance of self-defense as a general defense in criminal law. Self-defense allows individuals to protect some values using force, skirting the monopoly on power enjoyed by the state.

Being in a situation that warrants self-defense is considered to nullify the individual's fault required for the imposition of criminal liability. This concept has been accepted by legal systems worldwide since ancient times, and eventually the defense has become wider and more accurate.[84]

Its modern form allows individuals to repel attacks on legitimate interests. There are several conditions for this general defense to be applicable:

1. The protected interest must be legitimate. Legitimate interests are the life, freedom, body, and property[85] of the individual or of other individuals.[86] No previous introduction between the person threatened and the defender is required,[87] so that self-defense is not only of the "self."
2. The protected interest must be attacked illegitimately.[88] When a police officer, carrying a legal warrant, attacks an individual to make an arrest, the attack is considered legitimate.[89]
3. The protected interest must be in clear and imminent danger.[90]
4. The act of self-defense must be carried out to repel the attack, and it must be proportional to it,[91] and immediate.[92]
5. The defender must not be in a situation to control the attack or the conditions for its occurrence (*actio libera in causa*).[93]

If all these conditions are met, the individual is considered to be acting in self-defense, and therefore no criminal liability is imposed on him or her for the commission of the offense. But not every time an attack is repelled can it be considered self-defense; only when the act meets the listed conditions in full. The question is whether the general defense of self-defense is applicable to AI systems. The answer depends on the capabilities of the AI systems to meet the conditions just outlined, in the relevant situations.

At the heart of self-defense is the legitimate protected interest. When the protected interest is that of another person (a human interest protected by the AI system using self-defense), there seems to be no difficulty. Indeed, this is the broad basis for the legal activity of guard robots (of humans, prisoners, borders, dwellings, and so on). If these robots have the authority to repel attacks on human interests, self-defense is a legal justification. But the question is whether the self-interest of an AI system may also be legally protected by self-defense. Indeed, there are two questions here: one is moral, the other legal.

Morally, the question is whether we accept the idea of a robot protecting itself and having derivative rights, some of them constitutional.[94] The moral question has nothing to do with the legal one. Since the 1950s, the human approach to this moral question has generally been positive. Recall Asimov's third law, which states, "A robot must protect its own existence, as long as such protection does not conflict with the First or Second Laws."[95] According to this law, under the right circumstances the robot is not only authorized to protect itself, but it must do so. As already noted, however, this is not part of the legal question, and as far as the applicability of self-defense to robots is concerned, only the legal question is relevant.

The legal question is much simpler, and it asks whether an AI system has the ability to protect its own "life," freedom, body, and property. To protect these legitimate interests, the AI system must first possess them. Only if an AI system possesses property can it protect it; otherwise it can protect only the property of others. The same is true concerning life, freedom, and body. Currently, criminal law protects *human* life, freedom, and body. The question of whether a robot has analogous life, freedom, and body is one of legal interpretation, in addition to the moral questions involved.

Criminal law recognized decades ago that the corporation, which is not a human entity, possesses life, freedom, body, and property.[96] We return to this analogy in the next chapter when we discuss the punishments of AI systems, including capital punishment and imprisonment. But if the legal question concerning corporations, which are abstract creatures, has been decided affirmatively, it would be unreasonable to decide otherwise in the case of AI systems, which physically simulate these human attributes much better than do abstract corporations.

For example, in a prisoner escape, a prison-guard robot is attacked by the escaping prisoners, who intend to tie it up in order to incapacitate it and prevent it from interfering with their escape. Is it legitimate for the robot to defend its mission and protect its freedom? If the prisoners intend to break its arms, is it legitimate for the robot to defend its mission and protect its "body"? If the answers are analogous to those supplied for corporations, they are affirmative. In any case, even if the idea of an AI robot's life, body, and freedom is not acceptable, despite the analogy with corporations and despite the positive moral attitude, AI systems still meet the first condition of self-defense by protecting the life, body, freedom, and property of humans.

The other two legal conditions of self-defense do not appear to differ for humans and AI systems. The condition requiring an illegitimate attack depends on the attacker, not on the identity of defender, whether human or not. The condition of clear and imminent danger to the protected interest does not depend on the defender either. The nature of the danger is determined by the attacker, not by the defender, and therefore this condition is met not through the behavior of the defender, but through the analysis of the attack, independent of the defender's identity.

The condition of repelling the attack by a proportional and immediate action raises another question: is it legitimate for a robot to attack a human, even if the attack is in order to repel, by a proportional and immediate reaction, a human attack? This question raises two subsidiary questions. The first involves the legitimacy of preferring the defender's rights over those of the attacker, regardless of their identity as humans, corporations, or AI systems. The second concerns restriction on AI systems as defenders that differ from the restrictions imposed on humans and corporations. Both subsidiary questions have legal and moral aspects.

The first subsidiary question is generally answered with reference to risk taking by the attacker. Society creates peaceful social mechanisms for dispute resolutions, such as legal proceedings, arbitration, mediation, and so on. The attacker chooses not to use these peaceful mechanisms, but rather chooses to resort to illegal violence. When acting in this manner and bypassing the legal mechanisms of dispute resolution, the attacker takes the risk of triggering a response against the illegal attack. In this case, criminal law prefers the innocent reaction over the illegal action.[97] This answer recognizes no difference between various types of attackers and defenders.

There may be several moral answers to the second subsidiary question, but none is relevant to the legal answer. Therefore, despite Asimov's first law, noted earlier,[98] which prohibits robots from harming humans, robots are actually used in several ways that harm humans. Using robots in military and police functions (as soldiers, guards, armed drones, and so on)[99] inherently implies the possibility of causing harm to humans. As long as this use has not been explicitly prohibited by criminal law, it is legitimate. It appears that legal systems worldwide have not only made their legal choice in this matter, but have made their *moral* choice as well, rejecting Asimov's first law, which has been regarded as too panicky.

Consequently, criminal law sees no legal problem with the possibility that a robot may protect legitimate interests under illegal attack by humans. This question would not have been raised at all if the attacker were a robot, in which case the answer would be that the attack must be repelled with proportional and immediate action, whether the defender is human or not. The reaction is evaluated identically, regardless of whether the defender is human. *Repelling* means that the act is a reaction to the attack, which is the cause for the reaction. *Immediate* means that the reaction is subsequent to the attack. The reaction is not illegal revenge, but an act intended to neutralize the threat.

Proportional means that the defender does not use excessive force to neutralize the threat. The defender must evaluate the threat and the means that may be used to neutralize it, and then choose the means that are not excessive in relation to the specific threat under the given circumstances. Proportionality resembles reasonability in many ways, and some understand it as being part of reasonability. Therefore, to meet this condition, the AI system must have the capability of reasonableness, which is required for negligence as well, as discussed previously in that context.[100]

Finally, it is required that the defender have no control over the attack or the conditions for its occurrence. This requirement is intended to impose criminal liability on persons who brought the attack upon themselves and then attempt to save themselves from criminal liability by using the self-defense argument. The court is likely to seek out the deep reasons for the attack. If the defender was innocent in this context, the defense is applicable. This requirement is not applied differently to humans, corporations, or AI entities. As long as the AI entity was *not* part of such a plot, the general defense of self-defense is applicable to it. The proof may be based on the AI system's records. Consequently, it appears that the general defense of self-defense is applicable to AI systems.

5.3.2. Necessity

Can an AI system be considered to be acting out of necessity in the criminal law context? Necessity is a justification of the same type as self-defense. Both are designed to mitigate the society's absolute monopoly on power, discussed earlier.[101] The principal difference between self-defense and necessity is in the identity of the object of the response. In the case of self-defense, the defender's reaction is against the attacker, whereas in the case of necessity it is against an innocent object (innocent person, property, and so on). The innocent object is not necessarily related to the cause of the reaction.

For example, two people are sailing on the high seas. The boat hits an iceberg and sinks. Both people survive on an improvised raft, but have no water or food. After a few days, one of them eats the other in order to survive.[102] The victim did not attack the eater and was not to blame for the crash; she was completely innocent. The attacker was also innocent, but he knew that if he did not eat the other person, he would definitely die. If the eater is indicted for murder, he can claim necessity (self-defense is not relevant in this case because the victim did not threaten the attacker).

The traditional approach toward necessity is that under the right circumstances, it can justify the commission of offenses (*quod necessitas non habet legem*).[103] The traditional justification of the defense is the understanding in criminal law of the weakness of human nature. The individual who acts under necessity is considered to be choosing the *lesser of two evils* from his or her own point of view.[104] In our example, if the attacker chooses not to eat the other person, they both perish; if he eats the other person, only one of them dies. Both situations are evil, but the lesser evil of the two is the one in which one person survives. The victim of necessity is not blamed for anything but is innocent, and the act is considered to be justified.[105]

Given that the act performed out of necessity is the individual's, carried out when the authorities are not available, the general defense of necessity partially reduces the applicability of the general concept whereby only society (the state) has a monopoly on power, as we have discussed. Being in a situation that requires an act of necessity is considered to nullify the individual's fault required for the imposition of criminal liability. This concept has been accepted by legal systems worldwide since ancient times, and eventually the defense became broader and more accurate, so that its modern form allows individuals to protect legitimate interests.[106]

There are several conditions for this general defense to be applicable:

1. The protected interest must be legitimate. Legitimate interests are the life, freedom, body, and property of the individual or of other individuals. No previous introduction between the person threatened and the defender is required.[107]
2. The protected interest must be in clear and imminent danger.[108]
3. The act of necessity must be directed against an external and innocent interest.[109]
4. The act of necessity must be carried out to neutralize the danger, and it must be proportional to it,[110] and immediate.[111]
5. The defender must not be in a situation to control the danger or the conditions for its occurrence (*actio libera in causa*).

If all these conditions are met, the individual is considered to be acting under necessity, and therefore no criminal liability is imposed on him or her for the commission of the offense. But not every danger neutralized by causing harm to an innocent interest can be considered necessity; only when the act meets the listed conditions in full. The question is whether the general defense of necessity is applicable to AI systems. The answer to this question depends on the capabilities of the AI systems to meet the conditions just outlined, in the relevant situations.

Four of the listed conditions—numbers 1, 2, 4, and 5—are identical with those of self-defense, *mutatis mutandis*; the difference is that instead of an attack on a legitimate interest, we are facing danger to that same interest. The main difference between self-defense and necessity lies in one of the conditions. Whereas in self-defense the act is directed toward the attacker, in the case of necessity the act is directed toward an external or innocent interest. In the case of necessity, the defender must choose between the lesser of two evils, one of which harms an external interest, who may be an innocent person having nothing to do with the danger.

The question regarding AI systems in this context is whether they are capable of choosing the lesser of two evils. For example, an AI-based locomotive transporting twenty passengers arrives at a junction of two railroads. A child is playing on one of the rails, but the second rail leads to a cliff. If the system chooses the first rail, the child will certainly die but the twenty passengers will survive. If it chooses the second rail, the child will survive, but given the vehicle's speed and the distance from the cliff, it will certainly fall off the cliff and none of the passengers will survive.

If the vehicle had been driven by a human driver who chose the first rail and killed the child, no criminal liability would be imposed on him owing to the general defense of necessity. An AI system can calculate the probabilities for each option and choose the possibility with the lowest number of casualties. Strong AI systems are already used to predict complicated events (for example, weather patterns), and calculating the probabilities in the locomotive example is considered much easier. The AI system analyzing the possibilities is faced with the same two possibilities that the human driver has to consider.

If the AI system takes into account the number of probable casualties, based on its programming or as a result of the relevant machine learning, it will probably choose to sacrifice the child. This choice by the AI system meets all the conditions of necessity, so that if it were human, no criminal liability would be imposed on it owing to the general defense of necessity. Should the AI system be treated differently, then? Moreover, if the AI system chooses the alternative option, and causes not the lesser but the greater of the two evils to occur, we would probably want to impose criminal liability (on the programmer, the user, or the AI system) in exactly the same way as if the driver were human.

Naturally, some moral dilemmas may be involved in these choices and decisions: for example, whether it is legitimate for the AI system to make decisions in matters of human life, and whether it is legitimate for AI systems to cause human death or severe injury. But these dilemmas are no different from the moral dilemmas of self-defense, discussed earlier.[112] Moreover, moral questions are not to be taken into consideration in relation to the question of criminal liability. Consequently, it appears that the general defense of necessity is applicable to AI systems in a similar way to self-defense.

5.3.3. Duress

Can an AI system be considered as acting under duress in the criminal law context? Duress is a justification of the same type as self-defense and necessity. All are designed to mitigate the society's absolute monopoly on power, already discussed.[113] The principal difference between self-defense, necessity, and duress is in the course of conduct. In self-defense, the defender's reaction is intended to repel the attacker; in necessity, it is a reaction against an external innocent object; and in the case of duress, it is surrendering to a threat by committing an offense.

For example, a retired criminal with expertise in breaking into safes is no longer active. He is invited by ex-friends to participate in another robbery, where his expertise is required. He says no. They try to convince him, but he still refuses. So they kidnap his daughter and threaten him that if he does not participate in the robbery, they will kill her. He knows them well, and knows that they are capable of doing it. He also knows that if he involves police, they will kill his daughter. As a result, he surrenders to the threat, participates in the robbery, and uses his expertise. If captured, he can claim duress. Self-defense and necessity are irrelevant in this case because he surrendered to the threat rather than face it.

The traditional approach toward duress is that under the right circumstances, it can justify the commission of offenses.[114] The traditional justification of the defense is the understanding in criminal law of the weakness of human nature. At times, people would rather commit an offense under threat than face the threat and pay the price by causing harm to precious interests. Until the eighteenth century, the general defense of duress was applicable to all offenses.[115] Later, the Anglo-American legal systems narrowed its applicability, and today it no longer includes severe homicide offenses that require *mens rea*.[116]

In the safe-breaking example, if the offense were not robbery but murder, the general defense of duress would not be applicable for the imposition of criminal liability, but only as a consideration in punishment. The reason for the narrow applicability is the sanctity of human life.[117] But there are many exceptions to this approach.[118] In general, with the narrow exception for homicide, duress enjoys broad application as a general defense in criminal law worldwide. The individual who acts under duress is considered to be choosing the lesser of two evils, from his or her point of view: the evil of committing an offense by surrendering to the threat versus the evil of harming a legitimate interest.

Given that the act performed under duress is the individual's, carried out when the authorities are not available or effective, the general defense of duress partially reduces the applicability of the general concept whereby only society (the state) has a monopoly on power, as discussed earlier. Being in a situation that requires an act of duress is considered to nullify the individual's fault required for the imposition of criminal liability. The modern basis of this defense is the understanding that an individual may surrender to a threat rather than face it because not all people are heroes, and no one is required by law to be a hero.[119]

There are several conditions for this general defense to be applicable:

1. The protected interest must be legitimate. Legitimate interests are the life, freedom, body, and property of the individual or of other individuals. No previous introduction between the person threatened and the defender is required.[120]
2. The protected interest must be in clear and imminent danger.[121]
3. The act of duress must be in surrender to a threat.
4. The act of duress must be proportional to the danger.[122]
5. The defender must not be in a situation to control the danger or the conditions for its occurrence (*actio libera in causa*).[123]

If all these conditions are met, the individual is considered to be acting under duress, and therefore no criminal liability is imposed on him or her for the commission of the offense. But not every person surrendering to a threat can be considered to act under duress; only when the act meets the listed conditions in full. The question is whether the general defense of duress is applicable to AI systems. The answer depends on the capabilities of the AI systems to meet the conditions just outlined, in the relevant situations.

Four of the listed conditions—numbers 1, 2, 4, and 5—are almost identical with those of self-defense and necessity, *mutatis mutandis*; the difference is that in most legal systems, duress does not require an immediate act because the threat and danger to the legitimate interest may be continuous. The main difference between self-defense, necessity, and duress lies in one of the conditions. In self-defense, the act is directed toward the attacker; in the case of necessity, the act is directed toward an external or innocent interest; and acting under duress is surrender to a given threat.

The question concerning AI systems in this context is whether they are capable of choosing the lesser of two evils. For example, a prison-guard robot captures an escaping prisoner, who points a loaded gun at a human prison guard and threatens that if he is not released immediately by the robot, he will shoot the human guard. The robot calculates the probabilities and determines that the danger is real. If the robot surrenders to the threat, the guard's life is saved, but an offense is committed and the robot becomes an accessory to escape. If the robot does not surrender, no offense is committed, the escape is foiled, but the human guard is killed.

If the guard who captured the prisoner were human, then no criminal liability would be imposed on him owing to the general defense of duress because all the conditions of this defense are met. An AI system can calculate the probabilities for each option and choose the possibility with the lowest number of casualties. Strong AI systems are already used to predict complicated events (for example, weather patterns), and calculating the probabilities in the prison-guard example is considered much easier. The AI system analyzing the possibilities is faced with the same two possibilities that the human guard has to consider.

If the AI system takes into account the number of probable casualties, based on its programming or as a result of the relevant machine learning, it will probably choose to surrender to the threat. This choice by the AI system meets all the conditions of duress, so that if it were human, no criminal liability would be imposed on it owing to the general defense of duress. Should the AI system be treated differently, then? Moreover, if the AI system chooses the alternative option, and causes not the lesser but the greater of two evils to occur, we would probably want to impose criminal liability (on the programmer, the user, or the AI system) in exactly the same way as if the prison-guard robot were human.

Naturally, some moral dilemmas may be involved in these choices and decisions: for example, whether it is legitimate for the AI system to make decisions in matters of human life, and whether it is legitimate for AI systems to cause human death or severe injury. But these dilemmas are no different from the moral dilemmas of self-defense and necessity, discussed earlier.[124] Moreover, moral questions are not to be taken into consideration in relation to the question of criminal liability. Consequently, it appears that the general defense of duress is applicable to AI systems in a similar way to self-defense and necessity.

5.3.4. Superior Orders

Is an AI system on official duty, committing an offense under orders from its superior, protected from criminal liability? The general defense of superior orders is relevant for individuals who discharge official duties in authoritarian hierarchical official organizations such as the army, police, and rescue forces. These individuals are often required to act against their natural instincts. When faced with a gigantic fire, one's natural instinct is to escape, not to walk into it and save trapped people. One of the strongest factors in the success of such missions is discipline. Disciplined soldiers are likely to fulfill their mission even if this involves risking their lives. Therefore, the first thing that soldiers are taught is discipline.

But at times, orders from superiors contradict criminal law, and carrying them out involves committing an offense. There are two extreme models for a solution to this problem. The first is the *absolute defense model*, in which all actions under superior orders are protected.[125] This model places discipline above the rule of law, and only the commanders are to be held criminally liable according to this model. The other extreme model is the *absolute responsibility model*, in which no action is protected from criminal liability, even if performed under superior orders.[126] This model places the rule of law above discipline.

Under the first extreme model, the individual has no discretion in committing the offense. This model has led to the commission of war crimes and crimes against humanity during World War II and other wars. Under the second extreme model, the individual must be an expert in criminal law or accompanied by an attorney at all times to make sure of not being criminally liable for actions ordered by superiors. The disadvantages of these models have led to the emergence of moderate models, the most common of which is the *manifestly illegal order* model. According to this model, the individual is protected from criminal liability unless he or she performs a manifestly illegal order; if the individual performs an illegal order that is not *manifestly* illegal, he or she is protected.[127]

The ultimate question concerns the difference between illegal and *manifestly* illegal orders. Both types objectively contradict the law, but the manifestly illegal order contradicts the public policy as well. Every society follows a public policy that reflects its common values. These values may be moral, social, cultural, religious, and so on. Different societies embrace different values and have differing public policies. A manifestly illegal order is one that harms these values, and thus harms society's public policy and its self-image.

For example, a soldier is ordered to rape a civilian who resists the military operation. Rape is illegal, and in this situation it is manifestly illegal because it violates the basic values of modern Western society. There is no accurate and conclusive examination to distinguish between illegal and manifestly illegal orders because public policy is dynamic, and it changes with time, population, and social trends. But public policy may be taught, usually inductively, case to case. Consequently, two main conditions must be met for the general defense of superior orders to be applicable:

1. hierarchical subordination to authorized public authority; and
2. superior orders that require obedience and are not manifestly illegal.

If both conditions are met, the individual is considered to be acting under superior orders, and therefore no criminal liability is imposed on him or her for the commission of the offense. But not every time an individual obeys superior orders is the general defense applicable; only when the act meets the listed conditions in full. The question is whether the general defense of superior orders is applicable to AI systems. The answer to this question depends on the capabilities of the AI systems to meet the conditions just outlined, in the relevant situations.

The first condition has to do with the objective characteristics of the relationships between the individual and the relevant organization.[128] This relationship must involve a hierarchical subordination to authorized public authority and a legitimate and operative system of hierarchical order. Such systems exist in the army, police, and so on. Private organizations, however, have no authority to commit offenses. AI systems are in use in many of these organizations, including the military, police, and prisons, and are operated under superior orders for their regular activity. The systems perform various tasks under these orders.

The second condition has to do with the characteristics of the given superior order. The order must require obedience; otherwise, it cannot be considered an order. As far as its content is concerned, the order must not be manifestly illegal. If the order is legal or illegal, but not manifestly illegal, it satisfies this condition. Classification of the order as illegal or manifestly illegal is at the discretion of the court, but it may be learned inductively based on analysis of individual cases. AI systems equipped with machine learning facilities are capable of inferring at least the general outlines of manifestly illegal orders.[129]

For example, an AI drone operated by the air force is on a mission to seek out a terrorist lab and destroy it. The drone finds its target, and then delivers the information to headquarters. The information indicates that the lab is populated by a known terrorist and his family. The drone receives the order to attack the lab using a heavy bomb. After calculating the probabilities, the drone determines that all the people in the lab will die, and then it executes the order. When the records of the drone are examined, it emerges that the drone understood the legality of the orders to be in a grey area regarding the terrorist's family, and that it could have flown lower and destroyed the lab with fewer casualties.

If the drone were human, it would probably claim the general defense of superior order, because international law accepts such orders under certain situations, as long as the order is not manifestly illegal. Therefore, a human pilot would probably have been acquitted in such a case, and the general defense would have been applicable. Should an AI system be treated differently? If both the human pilot and the AI system have the same functional discretion, and both meet the conditions of this general defense, there is no legitimate reason for using a double standard in such cases.

Naturally, the criminal liability of the AI system, if any, does not affect the criminal liability of its superiors, if any (in case of an illegal order). Some moral dilemmas may also be involved in these choices and decisions: for example, whether it is legitimate for the AI system to make decisions in matters of human life, and whether it is legitimate for AI systems to cause human death or severe injury. But these dilemmas are no different from the moral dilemmas involving the applicability of other general defenses, discussed earlier.[130] Moreover, moral questions are not to be taken into consideration in relation to the question of criminal liability. Consequently, it appears that the general defense of superior orders is applicable to AI systems.

5.3.5. De Minimis

Can the general defense of *de minimis* be applicable for AI systems? In most legal systems, offenses are defined and formulated broadly, inevitably resulting in over-inclusion or over-criminalization, so that cases that are not supposed to be considered criminal end up being included within the scope of the offense. At times, the criminal proceedings in these cases would cause more social harm than benefit. For example, the case of a fourteen-year-old girl who steals her brother's basketball falls within the scope of the offense of theft, and the question is whether it would be appropriate to institute criminal proceedings for theft in such a case, considering its social consequences.

Most legal systems solve such problems by granting broad discretion to the prosecution and the courts. The prosecution may decide not to initiate criminal proceedings in cases of low public interest, and if such proceedings are started, the court may decide to acquit the defendant owing to low public interest. When the prosecution exercises this option, it is within its administrative discretion. When the court exercises this discretion, it is within its judicial power through the application of the general defense of *de minimis*, which enables the court to acquit the defendant based on low public interest in the case.

This type of judicial discretion has been widely accepted since ancient times. Roman law, for example, determined that criminal law does not extend to infant and petty matters (*de minimis non curat lex*), and that the judge should not be troubled with such matters (*de minimis non curat praetor*).[131] In modern criminal law, the general defense of *de minimis* is seldom exercised by the court given the wide administrative discretion of the prosecution. But in some extreme cases, the court may exercise this judicial discretion, in addition to the administrative discretion of the prosecution.[132]

The basic test for *de minimis* defense is one of the social endangerment involved in the commission of the offense. For the general defense of *de minimis* to be applicable, the offense should reflect extremely low social endangerment.[133] Naturally, different societies at various times may attribute different levels of social endangerment to the same offenses, because social endangerment is dynamically conceptualized through morality, culture, religion, and so on. The relevant social endangerment is determined by the court. The question is whether this general defense is relevant for offenses committed by AI systems.

Applicability of *de minimis* defense depends on the particulars of the case at hand and is based on its relevant aspects, and not necessarily on the identity of the offender. The personality of the offender may also be taken into consideration, but only as part of assessing the case. For this reason, there is no difference between humans, corporations, or AI entities regarding the applicability of this defense. The required low social endangerment is reflected in the factual event (*in rem*). For example, a human driver swerves on the road and hits the guardrail. No damages are caused to the guardrail or to other property, and there are no casualties. This is a case for the *de minimis* defense, although it may be within the scope of several traffic violations.

Would this case be legally different if the driver were not human but an AI entity? And would it be different if the car were owned by a corporation? There is no substantive difference between humans, corporations, and AI systems regarding the applicability of the *de minimis* defense, especially in light of the fact that this general defense is based on the characteristics of the factual event and not necessarily on those of the offender.

Consequently, it appears that the general defense of *de minimis* is applicable to AI systems.

CLOSING THE OPENING EXAMPLE: ROBOT KILLING IN SELF-DEFENSE

We opened this chapter with the example that described an AI robot, functioning as a prison guard, who killed in self-defense. Returning to this example, we can analyze it based on the insights gained from this chapter. Let us assume that after examining the robot's records, it transpires that the robot had consolidated awareness of the standards of criminal law regarding the factual components of the homicide. There is no doubt that the human prison guard died as a result of being smashed against the wall by the robot (factual causal connection). Following through this example step by step, we may find several applicable general defenses.

First, the virus planted in the main computer interfered with the robot's ability to create the right image of factual reality, and caused it to ignore some of the rules of the system. Two types of general defenses may be applicable as a result. The first is factual mistake, discussed previously. The reason for the factual mistake is immaterial for its legal consequences, and therefore, if during the event the factual image of the AI system differed from what actually happened because of the effect of the virus, this can serve as the basis for the applicability of factual mistake general defense.

The effect of the virus on the cognitive or volitive malfunction of the robot during its operation should also be examined based on the records of the AI system. If the virus caused the malfunction, then this can be the basis for the applicability of another general defense having to do with loss of self-control, insanity, or intoxication. The choice between these options depends on the concrete effect of the virus on the AI robot. Because the planting of the virus was involuntary (done by hackers), the consequences of the effects of the virus will most likely prevent imposition of criminal liability if the conditions of the relevant defense are met.

When the robot apprehended the two struggling humans, it misinterpreted the situation as one in which two prisoners were struggling. This misinterpretation can be proven by analyzing the robot's records. If the robot indeed misinterpreted the situation, whether or not as a result of the virus, the general defense of factual mistake is applicable. The same is true for the robot misinterpreting the command to keep away as if this had been a threat to its mission. If the robot indeed interpreted this command as a threat, whether or not it was a threat in reality, can the reaction of grabbing the human guard and smashing him against the wall be justified and defended?

As understood by the robot, this reaction is at most one of repelling a threat. If all the conditions of self-defense are met, this may serve as the basis for the general defense of self-defense. The robot's mission, however, was to prevent the escape of other prisoners but not to protect their lives, freedom, body, or property. Therefore, self-defense is inapplicable in this case, as are necessity and duress. Even if the system had orders to kill resisting prisoners, the general defense of superior orders would not have been applicable. Killing prisoners in response to an attempted escape, that results in the escape of other prisoners (the disguised prisoner managed to escape as a result of the robot's actions), would likely be considered as following a manifestly illegal order, obeying which is not protected.

But if it can be proven, based on the records, that the robot's decision to kill the guard was a direct consequence of the effect of the virus on the system, then the legal situation may be different. If the virus caused the robot to act automatically when killing the guard, or caused a malfunction in its cognitive or volitive processes as a result of which the robot killed the guard, then the general defenses of loss of self-control, insanity, or intoxication may be applicable according to the concrete effect of the virus.

Naturally, this does not reduce the criminal liability of others involved in the event, including humans (superiors, prisoners, guards, and hackers), corporations (the prison service and the manufacturer of the AI systems), and other AI systems. It appears, however, that general defenses in criminal law are applicable to AI systems as well as to humans and corporations.

6

SENTENCING AI

Is it possible to impose a prison sentence on an AI robot? How can such a punishment be carried out in practice? One of the objectives of the criminal process is sentencing. The legal discussion taking place in court regarding the criminal liability of the defendant is often considered to be a preliminary discussion before sentencing. Debating the criminal liability of the defendant is not a theoretical exercise, but it has practical consequences. If the defendant is acquitted, no punishment is imposed. But if he or she is convicted, the punishment must reflect the attitude of society toward the commission of the offense and toward the personal circumstances of the defendant.

The question is whether the sentencing process in criminal law and criminal punishments are relevant to machines. For example, after all the factual and mental element requirements of manslaughter were met by an AI robot, the court convicted the robot of manslaughter, and the prosecution and defense presented evidence for punishment. Under the circumstances, if the robot had been a human, the court would have sentenced this offender to three years of imprisonment. Is such a punishment applicable to a robot? Is criminal punishment applicable to machines at all?

This question is relevant and complementary to the examples that began each of the previous chapters. To answer it, we must first explore the general purposes of sentencing and their applicability to AI entities, and then understand the legal technique of imposing punishments on non-human entities. Finally, we will discuss the applicability of punishing AI entities.

6.1. CONCEPTUAL APPLICABILITY OF CRIMINAL PUNISHMENT OF AI ENTITIES

6.1.1. AI Purpose of Sentencing: Combining Rehabilitation and Incapacitation

Criminal punishments are imposed for a purpose. There are four common general purposes of sentencing, which also function as the general sentencing considerations: retribution, deterrence, rehabilitation, and incapacitation.

Retribution (*lex talionis*) is the most ancient purpose of sentencing.[1] It has to do with the past, and contains many aspects of revenge. The basic idea of retribution is expressed in the maxim "an eye for an eye,"[2] that is, whatever you have caused shall be imposed on you. Retribution purposefully causes suffering to the offender to a degree that fits the suffering caused by the offender, and not more. Retribution functioned as a lenient form of punishment because it prevented society from causing greater suffering to offenders than they have caused to others.[3] For example, the death penalty would not be appropriate for breaking someone's tooth.

The importance of retribution as a major objective of punishment gradually diminished between the 1920s and the 1970s, primarily because of criticism that retribution was archaic and retrospective, and therefore did not sufficiently take into account the social benefit of punishment.[4] The purpose of retribution was considered barbaric.[5] Nevertheless, in light of the actual and conceptual failure of rehabilitation, as we will discuss, since the 1970s retribution has regained a significant position among the purposes of punishment, in its modern form of "just desert."[6]

Whereas rehabilitation concentrates on the personal characteristics of the offender, thereby punishing the offender for his or her character, retribution concentrates on the characteristics of the offense.[7] The "just desert" approach incorporates some personal aspects of the offender into the retribution, thereby softening it.[8] Thus, there are two important factors in evaluating the offender's punishment under retribution: (1) the factual damage caused by the offender to society, and (2) the personal culpability of the offender.[9] Personal culpability is measured mostly by the mental element of the offender.[10]

The question for our discussion is whether retribution is relevant to AI systems. Retribution is intended primarily to satisfy society. Causing suffering to the offender, in itself, has no prospective value. The suffering may deter the offender, but that is part of the general purpose of deterrence, not of retribution. Retribution may have a cathartic effect on society and the victims by causing the offender to suffer, but in this context, punishing machines by retribution would be meaningless and impractical.

Some people, when they are in a great hurry and their car will not start, may become angry and hit the car, kick it, or yell at it. Punishing machines, including highly sophisticated AI robots, by retribution would be the same as kicking a car. Some people may achieve a certain release of their anger, but nothing more would be accomplished. Machines do not suffer, and as long as retribution is based on suffering, it is not relevant to punishing robots. This is true for both classic and modern ("just deserts") approaches to retribution.

Further, if retribution functions as a lenient form of sentencing aimed at preventing revenge, its irrelevance to AI sentencing is even more pronounced. Revenge is assumed to cause more suffering to the offender than would the official punishment, but because machines do not experience suffering, the distinction between revenge and retribution is meaningless for them.

Deterrence is the teleological response to deontological retribution. Deterrence emerged as the general purpose of sentencing toward the end of the eighteenth century. Inspired by utilitarianism, legal scholars searched for the prospective value of the law. This search included sentencing as well, and it was assumed that criminal punishment may discourage offenders and potential offenders from committing further offenses. Some scholars called for the inclusion of punishment components as integral parts of the definition of offenses.[11] Deterrence reflected the human wish for a better society, without delinquency.[12]

This better society was supposed to be achieved by intimidation. Offenders and potential offenders, afraid of punishment, were expected to reconsider their delinquent thoughts and eventually renounce their criminal plan. Because punishment was supposed to deter people from committing offenses, it was assumed that the more severe the punishment, the more people would avoid committing offenses. But this assumption turned out to be misleading for two reasons.

First, people are deterred more by efficient enforcement than by punishment.[13] For example, the effect on driving habits of doubling the maximum penalty from $100 to $200 for exceeding the speed limit on a road that has practically no police presence would be insignificant. Yet if more police officers patrolled the same road, even if the fine were lowered to $80, drivers would be more deterred from speeding. When city inspectors are on strike and no citations are issued for parking in restricted areas, most people tend to park their cars in these areas, regardless of the level of maximum punishment. Thus, the deterrent value of punishment alone is very limited.[14]

Second, the deterrent value of punishment, even if it is maximal, tends to erode in time with use, and at some point, the punishment may end up functioning as an incentive for repeated offense.[15] For instance, when an offender is sentenced to prison for the first time, the deterrent effect of the punishment is at its maximum level for that offender. The second time, the effect is reduced, and the tenth time it is reduced dramatically. After spending forty years in prison, the prisoner does not really have a life outside of prison. Prison provides meals, social status, social security, and friends. If released, the prisoner may commit offenses only to return to this familiar social environment.

Thousands of homeless people in the Western world are committing offenses on cold days to find shelter from the cold, have a warm meal, and find a place to sleep. Not only does punishment *not* serve as a deterrent, but it serves as an incentive to repeat offending. In other words, the punishment turns into the benefit of the offense. Nevertheless, for most offenders and potential offenders, a combination of proper enforcement with appropriate punishments can supply some deterrence, even if this is not entirely effective. Our question, then, is whether deterrence is relevant to AI systems.

Deterrence is intended to prevent the commission of the next offense, by using intimidation. In the current technology, intimidation is a feeling that machines cannot experience. Intimidation is based on a fear of future suffering imposed if an offense is committed. Because machines do not experience suffering, as we have already noted, both intimidation and the reason for it are nullified when considering appropriate punishment for robots. At the same time, both retribution and deterrence may be relevant purposes for the punishment of human participants in the commission of offenses by AI entities (for example, users and programmers).

Rehabilitation is a relatively newer purpose of modern sentencing, based on the idea that the punishment and the process of sentencing may serve as opportunities for correcting offenders' social problems. Rehabilitation was inspired by the religious concept that every person may be corrected given the right conditions, and has been embraced as a major sentencing consideration since 1895 in Britain, spreading to the rest of the world.[16] The golden age of rehabilitation was between the 1920s and 1970s,[17] and its popularity rested on its prospective quality that offered hope for a better world.

Enthusiasm for rehabilitation waned with the publication of the Martinson report in 1974.[18] The publication examined rehabilitation reports in the state of New York and came to the surprising conclusion that "nothing works."[19] The conclusions were reexamined repeatedly, and each time confirmed,[20] giving rise to various criticisms of rehabilitation, on the part of research and even popular art.[21] Disappointment with rehabilitation triggered a revival of retribution, in its modern form as "just desert," as already noted. Rehabilitation was not abolished entirely, however, and it is still being considered in the sentencing process, but more carefully.

The prospective purpose of rehabilitation is to address the roots of delinquency. By solving the problems that lead to it, delinquency may be solved, or at least dramatically reduced. To this end, the court must understand the causes of an offender's delinquency, assess the potential for rehabilitation (using such professionals as social workers and psychologists), and eventually decide on a proper punishment aimed at solving the delinquency problem.[22] One of the most popular trends in rehabilitation is to grant the offender the cognitive and social skills needed to cope with daily life.[23]

These tools are supposed to cause the offender to make inner changes in relation to his or her social, family, or professional environment, providing an opportunity for resocialization and reintegration into society and for leaving delinquency behind, to benefit the offender and also society.[24] Our question is whether rehabilitation is relevant to AI systems.

AI systems can experience decision-making processes and reach decisions that may appear unreasonable, as we have discussed.[25] At times, an AI system may need external guidance in order to refine its decision-making process, which may be part of the machine-learning process. For AI entities, rehabilitation functions in exactly the same way as for humans, causing them to make better decisions in their everyday lives from the point of view of society. Following this approach, the punishment of AI entities would be directed at refining the machine-learning process.

Having been rehabilitated, an AI system would be able to form better and more accurate decisions, by applying more limitations to its discretion and by refining the process through machine learning. Thus, the punishment, if correctly applied to an individual AI system, would become part of the machine-learning process. Directed by the rehabilitative punishment, the AI system would therefore have better tools to analyze factual data. This is the same effect that rehabilitative punishment is intended to have on humans by granting them better tools for facing factual reality.

Consequently, rehabilitation can be a relevant purpose of punishment for AI systems because it is not based on intimidation or suffering, but is directed at improving the performance of the AI system. For humans, this consideration may be secondary in many cases, but for AI systems it may be a primary purpose of punishment. Nevertheless, rehabilitation is only one of the considerations that are relevant to AI systems; incapacitation is another.

Incapacitation is the physical prevention of offenders from committing further offenses. This general consideration is based on skepticism toward the effectiveness of the other prospective considerations (deterrence and rehabilitation) to reduce delinquency rates; it focuses therefore on the physical capabilities of delinquency. The basic approach is to neutralize the offender's physical capabilities to commit offenses.[26] For example, incapacitation is the general consideration that justifies the chemical castration of sex offenders by eliminating their sexual impulse and their ability to commit further sexual offenses.

For the physical prevention of delinquency, it is not necessary to bring about inner changes or cognitive corrections in the offender's mind. The offender is not required to change his or her personality or values for this consideration to be effective.[27] In some ways, the consideration of incapacitation is the expression of disappointment, even despair with human nature, contrary to rehabilitation, which expresses hope. When society feels hopeless toward an offender's willingness or mental capability to stop the delinquent activity, both deterrence and rehabilitation may appear to be ineffective; but further offense must still be prevented, so society resorts to incapacitation.

In this way, regardless of whether the offender understands or agrees with the demands of society, the next offense is prevented because the physical capability to commit the offense is neutralized. If in most cases, rehabilitation is considered a lenient general consideration of punishment, incapacitation is considered a harsh one. Our question is whether incapacitation is relevant to AI systems.

If an AI system commits offenses and lacks the capability to change its ways through machine learning, only incapacitation can supply an adequate answer. Regardless of whether the AI system understands the meaning of its activity, or whether the AI system is equipped with proper tools to perform inner changes, delinquency must still be prevented. In a situation of this type, society must disable the physical capabilities of the AI system to commit further offenses, despite its other skills. This is what the society does with human offenders in similar cases.[28]

We can conclude that two relevant considerations for the punishment of AI systems are rehabilitation and incapacitation. Both reflect extreme poles of sentencing, and both serve the purposes of criminal law regarding nonhuman offenders. When the AI system is capable of carrying out inner changes that affect its activity, rehabilitation appears to be preferable to incapacitation, but when it does not possess such capabilities, incapacitation is preferable. Thus, the punishment is adjusted to the relevant personal characteristics of the offender, as in the case of human offenders.

6.1.2. The Legal Technique of Conversion: The Case of Corporations (Round 2)

We have seen that sentencing considerations are relevant to AI systems. The question is, how can we impose human punishments on machines? For example, how can we impose imprisonment, fines, or capital punishment on AI systems? To do so, we need a legal technique of converting human penalties into penalties for AI entities. Such technique may be inspired by the one used to convert human penalties into penalties suitable for corporations. The first time we examined corporations (Round 1), we were concerned with the idea of imposing criminal liability on AI systems.[29] This time (Round 2) we are concerned with the idea of imposing punishments on these systems.

Not only has criminal liability been imposed on corporations for centuries, but corporations have also been sentenced, and not only to fines. Corporations are punished in various ways, including imprisonment. Note that corporations are punished separately from their human officers (directors, managers, employees, and so on) in exactly the way that criminal liability is imposed on them separately from the criminal liability, if any, of their human officers. There is no debate over the question whether corporations should be punished using a variety of punishments, including imprisonment—the question concerns only *how* to do it.[30]

To answer this question of "how," a general legal technique of conversion is needed, as just noted. This operation is carried out in three principal stages. First, the general punishment itself (for example, imprisonment, fine, probation, or death) is analyzed regarding its roots of meaning. Second, these roots are sought in the corporation. Third, the punishment is adjusted according to the roots found in the corporation. For example, in the case of imposition of incarceration on corporations, first incarceration is traced back to its roots in the act of depriving individuals of their freedom, and then a meaning is sought for the concept of freedom for corporations.

After this meaning has been understood, in the third and final stage, the court imposes a punishment that is the equivalent of depriving a corporation of its freedom. This is how the general legal technique of conversion works in the sentencing of corporations. At times, this requires the court to be creative in the adjustments needed to make punishments applicable to corporations, but the general framework is clear and workable, and it has been implemented with all types of punishments imposed on all types of corporations.[31]

An excellent example is the case of the Allegheny Bottling Company,[32] a corporation found guilty of price-fixing (antitrust). It was agreed that under the given circumstances, if the defendant were human, the appropriate punishment would be imprisonment for a certain term. The question concerned the applicability of imprisonment to corporations—in other words, a question of "how." As a general principle, the court declared that it "does not expect a corporation to have consciousness, but it does expect it to be ethical and abide by the law."[33]

The court did not find any substantive difference between humans and corporations in this matter, adding that "[t]his court will deal with this company no less severely than it will deal with any individual who similarly disregards the law."[34] This statement reflects the basic principle of equalizing punishments of human and corporate defendants.[35] In this case, the corporation was sentenced to three years' imprisonment, a fine of $1 million, and probation for a period of three years. The court proceeded to discuss the idea of corporate imprisonment based on the three stages just described.

First, the court asked what were the general meanings of imprisonment, and accepted the definitions of imprisonment as "constraint of a person either by force or by such other coercion as restrains him within limits against his will" and as "forcible restraint of a person against his will." The court's conclusion was simple and clear: "[t]he key to corporate imprisonment is this: imprisonment simply means *restraint*" and "*restraint*, that is, a deprivation of liberty." The court's conclusion was reinforced by several provisions of the law and also of case laws. Consequently, "[t]here is imprisonment when a person is under house arrest, for example, where a person has an electronic device which sends an alarm if the person leaves his own house."

This ended the first stage. In the second stage, the court searched for a meaning of this punishment for corporations, concluded that "[c]orporate imprisonment requires only that the Court restrain or immobilize the corporation,"[36] and proceeded to implement the prison sentence on the corporation according to this insight. Thus, in the third and final stage, the court made imprisonment applicable to the corporations and implemented it as follows:

> Such restraint of individuals is accomplished by, for example, placing them in the custody of the United States Marshal. Likewise, corporate imprisonment can be accomplished by simply placing the corporation in the custody of the United States Marshal. The United States Marshal would restrain the corporation by seizing the corporation's physical assets or part of the assets or restricting its actions or liberty in a particular manner. When this sentence was contemplated, the United States Marshal for the Eastern District of Virginia, Roger Ray, was contacted. When asked if he could imprison Allegheny Pepsi, he stated that he could. He stated that he restrained corporations regularly for bankruptcy court. He stated that he could close the physical plant itself and guard it. He further stated that he could allow employees to come and go and limit certain actions or sales if that is what the Court imposes.
>
> Richard Lovelace said some three hundred years ago, "stone walls do not a prison make, nor iron bars a cage." It is certainly true that we erect our own walls or barriers that restrain ourselves. Any person may be imprisoned if capable of being restrained in some fashion or in some way, regardless of who imposes it. Who am I to say that imprisonment is impossible when the keeper indicates that it can physically be done? Obviously, one can restrain a corporation. If so, why should it be more privileged than an individual citizen? There is no reason, and accordingly, a corporation should not be more privileged.
>
> Cases in the past have *assumed* that corporations cannot be imprisoned, without any cited authority for that proposition. . . . This Court, however, has been unable to find any case which actually held that corporate imprisonment is illegal, unconstitutional or impossible. Considerable confusion regarding the ability of courts to order a corporation imprisoned has been caused by courts mistakenly thinking that imprisonment necessarily involves incarceration in jail. . . . But since imprisonment of a corporation does not necessarily involve incarceration, there is no reason to continue the assumption, which has lingered in the legal system unexamined and without support, that a corporation cannot be imprisoned. Since the Marshal can restrain the corporation's lib-

erty and has done so in bankruptcy cases, there is no reason that he cannot do so in this case as he himself has so stated prior to the imposition of this sentence.[37]

Thus, imprisonment can be applied not only to human but also to corporate offenders. Following the same approach, imprisonment is not the only penalty applicable to corporations, but other penalties can be converted as well, even if they were originally designed for human offenders. And if this is true for imprisonment, which is an essentially human penalty, fine can be easily collected from corporations, just like taxes. This raises the equivalent question regarding AI systems.

Using the general legal technique of conversion, as just presented and applied to corporate delinquency, human punishments can be made applicable in the same way and imposed on AI systems as well. Next we examine the applicability of common punishments in modern criminal law to AI systems.

6.2. APPLICABILITY OF PUNISHMENTS TO AI ENTITIES

We will review the applicability and imposition of the following punishments on AI systems: capital punishment, imprisonment, probation, public service, and fine.

6.2.1. Capital Punishment

The death penalty is one of the most ancient punishments in human history, and it is considered the most severe penalty in most cultures.[38] In the past it was a common penalty, but since the eighteenth century, the global approach has been to restrict and minimize its use. Therefore, capital punishment has been replaced by more lenient penalties for many offenses, and methods of execution were developed to cause minimal suffering to the offenders.[39] For example, in the eighteenth century, the length of the noose was increased to cause a faster and less painful death, and the guillotine was introduced for the same reason.

Cruel methods of execution were prohibited. For example, dismembering by tying the offender's arms and legs to running horses was prohibited because of the suffering it caused. Some countries abolished capital punishment altogether, but most countries did not. The United States Supreme Court ruled in 1979 that capital punishment, in the appropriate cases, does not infringe on the Eighth Amendment to the Constitution (that is, it is not cruel and unusual punishment), and therefore it is constitutionally valid.[40]

The methods of execution used in the United States were not considered unconstitutional either.[41]

Retribution justifies capital penalty only for severe offenses in which penalty parallels the offense in suffering or result (for example, homicide offenses).[42] Deterrence can justify capital punishment only as a deterrent of the public and not of the offender, because a dead person cannot be deterred.[43] Rehabilitation is completely irrelevant for this punishment, as dead people cannot be rehabilitated. And given that retribution and deterrence are irrelevant for AI sentencing, as noted previously, and because rehabilitation is irrelevant to capital punishment, the only general consideration that supports capital punishment in the sentencing of AI entities is incapacitation.

There is no doubt that a dead person is incapacitated from committing further offenses and that the most dominant consideration for the death penalty is incapacitation.[44] The question regarding the applicability of capital punishment to AI systems is how to impose the death penalty on them. To answer this question, we must follow the three-stage model presented earlier.

First, we should analyze the meaning of capital punishment; second, seek out its roots as far as AI systems are concerned; and finally, adjust the punishment to these roots. Functionally, capital punishment is deprivation of life. Although this deprivation may affect not only the offender but other people as well (relatives, employees, and so on), the essence of the death penalty is the death of the offender. When the offender is human, life means the person's existence as a functioning creature. When the offender is a corporation or an AI system, its life may be defined through its activity.

A living AI system is a functioning one, therefore the "life" of an AI system is its capability to function as such. Stopping the AI system's activity does not necessarily mean the "death" of the system. Death means *permanent* incapacitation of the system's life. Therefore, capital punishment for an AI system means its *permanent* shutdown so that no further offenses or any other activity can be expected on the part of the system. When an AI system is shut down by court order, this means that the society prohibits the operation of that entity because it is considered too dangerous for society.

This application of capital punishment on AI systems serves both the purposes of capital punishment and of incapacitation (as a general purpose of sentencing) in relation to AI systems. When the offender is too dangerous for society, and society decides to impose the death penalty, if the punishment is acceptable in the given legal system, it is intended to accomplish the total and final incapacitation of the offender. This is true of human offenders, corporations, and AI systems. For AI systems, permanent incapacitation means absolute shutdown under court order, with no option of reactivating the system again.

A permanently incapacitated system will not be involved in further delinquent events. It may be argued that such shutdown may affect other innocent persons (for example, the manufacturer of the system, or its programmers and users). Yet this is true not only for AI systems, but for human offenders and corporations as well. The execution of an offender also affects his innocent family (in the case of a human offender) or affects employees, directors, managers, shareholders, and so on (in the case of a corporation). When the offender is an AI system, shutdown also affects other innocent persons, but this is not unique to AI systems. Therefore, capital punishment may be applicable to AI systems.

6.2.2. Imprisonment and Suspended Imprisonment

Imprisonment is a general term for various penalties. The common characteristic of these penalties is deprivation of the offender's liberty. Physical incarceration is only one type of imprisonment, although it is the most common. For example, under certain circumstances, public service can function as imprisonment. The purposes of imprisonment have differed in different societies and times. At times, the purpose was to make the offender suffer; at other times, it was to rehabilitate the offender under tight discipline.[45] But since the eighteenth century, a dominant consideration for the evaluation of imprisonment has been its social efficiency.[46]

The social efficiency of the imprisonment is evaluated by the rate of recidivism. If the rate of recidivism remains constant or increases, then imprisonment is not considered socially efficient. Therefore, the imposition of imprisonment has been used to initiate an inner change in the offender for the benefit of society.[47] Thus, prisoners have been taught professions; have learned reading and writing; have been helped to break drugs, alcohol, and violent habits; and have been encouraged to experience working during their stay in prison. Imprisonment has been adapted to fit various populations, resulting in the creation of such solutions as supermax prisons[48] and shock incarceration.[49]

Nevertheless, because imprisonment has become a popular penalty, prison overcrowding is currently considered its most acute problem. Retribution supports imprisonment, especially incarceration, because it makes the offender suffer. Deterrence supports imprisonment because the suffering caused in prison may deter offenders from recidivism and potential offenders from offending. Both retribution and deterrence, however, are irrelevant to AI systems because they experience neither suffering nor fear.⁵⁰ Imprisonment for AI systems should be evaluated based on rehabilitation and incapacitation, which are relevant to AI entities.

When the offender is deprived of his or her liberty, society can use this prison term for the offender's rehabilitation, by initiating inner change. This change can be the result of activity carried out in prison. If the offender accepts the inner change and does not return to delinquency, then imprisonment is considered successful and the offender rehabilitated. At the same time, when the offender is under strict supervision inside the prison, his or her capability for committing further offenses is dramatically reduced, which may be considered incapacitation as long as supervision prevents recidivism.

Here is the question regarding the applicability of imprisonment: how can imprisonment be imposed on AI systems? To answer this question, we again follow the three-stage model. First, we analyze the roots of the meaning of imprisonment; second, we seek the meaning of these roots for AI systems; and finally, we adjust the punishment to the roots we have revealed for AI systems.

Functionally, imprisonment is deprivation of liberty. Although this deprivation may affect not only the imprisoned offender, but other people as well (for example, relatives, employees, and so on), the essence of the imprisonment consists of restricting the offender's activity. When the offender is human, liberty means the person's freedom to act in any desired way. When the offender is a corporation or an AI system, its liberty may also be defined through its activity. The freedom of an AI system lies in the exercise of its capabilities without restriction, including both the exercise of these capabilities and their content.

Thus, the imposition of imprisonment on AI systems consists of depriving them of their liberty to act by restricting their activities for a specified period of time and under strict supervision. During this time the AI system may be repaired to prevent the commission of further offenses. Repair of the AI system may be more efficient if the system is incapacitated, especially if it is done under court order. This situation can serve both purposes of rehabilitation and incapacitation, which are the relevant sentencing purposes for AI systems. When the AI system is in custody, under restriction and supervision, its capability to offend is incapacitated.

An AI system being repaired through inner changes, initiated by external factors (for example, programmers working under court order), and gaining experience during the period of restriction, can be substantively considered to be in rehabilitation because the chances of the system being involved in further delinquency are being reduced. The social value of imposing imprisonment on AI systems is real, as the dangerous system is being removed from society for repairs, and at the same time, it is incapable of causing further harm to society. When the process is complete, the system may be returned to full activity. If the system is beyond rehabilitation, then incapacitation becomes the principal objective, dictating a long period of imprisonment or even capital punishment.

A sentence of suspended imprisonment is a conditional penalty.[51] The offender is warned that if further offense is committed, the full penalty of imprisonment is imposed for the new offense, and in addition, the offender will have to serve another term of imprisonment for the original offense. This penalty is intended to deter the offender from offending again, for at least as long as the condition remains in force. This penalty is imposed by adding the relevant line to the offender's criminal record. The question of suspended sentences does not concern the execution of the penalty, but concerns its social meaning when imposed on AI systems.

A sentence of suspended imprisonment for an AI system is an alert to reconsider its course of conduct. This process may be initiated by programmers, users, or the manufacturer, in the same way that human offenders may be assisted by their relatives or professionals (for example, psychologists or social workers), and corporate offenders by their officers or professionals. A suspended sentence is a relatively lenient measure calling for reconsideration of one's course of conduct. In essence, it is not substantially different from imprisonment, although for human offenders it may be vastly different because it spares the human offender the suffering of incarceration and deters him or her from delinquency through intimidation.

6.2.3. Probation

Prison overcrowding has forced states to develop substitutes for incarceration. One of the popular substitutes is probation, developed in the mid-nineteenth century by private charity and religious organizations that undertook to care for convicted offenders by giving them the social tools they needed to abandon delinquency and reintegrate into society.[52] Most countries embraced probation as part of their public sentencing systems. The first US state to do that was Massachusetts, in 1878. Thus, probation was operated and supervised by the state.[53]

The social tools provided to the offender vary from case to case. When the offender is addicted to drugs, the social tools include a drug rehabilitation program. When the offender is unemployed or has no profession, the social tools include vocational training. During the time the offender is under probation, the authorities supervise and monitor him or her to ensure that no further offenses are committed. Probation is dominantly rehabilitative, and is appropriate for offenders who have a high potential for rehabilitation. Consequently, the court needs an accurate assessment of that potential, prepared by the probation service, before sentencing the offender to probation.[54]

Retribution is irrelevant for probation, because probation is not intended to make the offender suffer. Neither is deterrence relevant to probation, which is perceived as a lenient penalty with a negligible deterrent value. Moreover, as we have already seen, both retribution and deterrence are irrelevant to AI sentencing. Incapacitation, which *is* relevant to AI sentencing, is not reflected in probation, unless the framework for the specific probation is extremely strict and the offender's delinquent capabilities are incapacitated. The dominant purpose of probation, however, is rehabilitation through the provision of the relevant social skills needed to reintegrate the offender into society.

Here is the question regarding the applicability of probation: how can probation be imposed on AI systems? To answer this question, we again follow the three-stage model. First, we analyze the roots of the meaning of probation; second, we seek the meaning of these roots for AI systems; and finally, we adjust the punishment to the roots we have revealed for AI systems.

Functionally, probation consists of supervising the offender and providing him or her with the means to reintegrate into society. These measures should match the type of delinquency that was the immediate cause for the sentencing.[55] Probation works as a functional correction of the offender. When an offense is committed by an AI system, the system must be diagnosed to determine whether it can be corrected. At this stage, human offenders are diagnosed to determine their potential for rehabilitation. Both types of diagnosis are performed by professionals.

Human offenders may be diagnosed by probation service staff, social workers, psychologists, psychiatrists, physicians, and so on. AI system offenders may be diagnosed by technology experts. If the diagnosis shows no potential for rehabilitation, then the ultimate purpose of sentencing becomes incapacitation, because society wishes to prevent further harm. But if the offender has a high potential for rehabilitation, then probation and rehabilitation as the purpose of sentencing are taken into consideration. This is equally true for human offenders, corporations, and

AI systems. The core question for AI system diagnosis is whether the system may be repaired in its current configuration, for example, through machine learning.

If the conclusion is that the appropriate penalty is probation, it must be applied with reference to the problems that the delinquency raised. This is true for all types of offenders. The objective of the treatment is to fix a certain problem, therefore the treatment should match the problem. The difference between probation and imprisonment, which also includes treatment, is the incapacitation of the offender's delinquent capabilities during the period of the penalty. If the offender is not considered dangerous during the period of the treatment, probation may be suitable, but if the offender must be incapacitated during treatment, imprisonment may be better suited.

Therefore, if society finds that the AI system can continue functioning while under treatment, probation may be suitable. If it is too dangerous for society, imprisonment may be suitable. When probation is finally imposed, the AI system must begin the treatment, which consists of repair of its inner processes. Some systems may require intervention in the machine-learning process, others require upgrades of their hardware, and still others require intervention in the basic software of the system. During this process, the AI system continues its routine activity under supervision imposed by the court.

Socially and functionally, probation is identical for human offenders, corporations, and AI systems. The attributes of each offender require different treatment, but that is true with respect to human offenders as well. Naturally, the manufacturers, programmers, and users of the system can initiate the repair process without the intervention of the court, but when the court orders the intervention, it means that this is the will of society. In the same way, a drug addict or the addict's family may initiate a drug rehabilitation process without a court order, but when it is imposed by the court, it signifies that this is the will of society.

6.2.4. Public Service

Public or community service is a substitute for imprisonment developed to ease prison overcrowding. For offenses that are not severe, the court may impose public service on the offender instead of other penalties, or in addition to them.[56] The offender is not incapacitated by this penalty, but on the contrary is forced to contribute to society in "compensation" for his or her involvement in the delinquency. In this way, society signifies that the delinquency is unacceptable, but because the social harm is not severe, lenient measures are taken to bring about the required inner change in the offender.

Public service has yet another dimension, which relates to the community. Public service is carried out in the offender's community to signify that he or she is part of that community, and that causing harm to the community boomerangs back onto the offender.[57] In many cases, public service is added to probation in order to improve the chances of the offender's full rehabilitation within the community.[58] Public service has more than mere compensational value, as it proposes to make the offender understand and become sensitive to the needs of the community. Public service is part of the learning and reintegration processes that the offender experiences.

Retribution is irrelevant for public service, because public service is not intended to make the offender suffer. Neither is deterrence relevant for public service because it is perceived as a lenient penalty with negligible deterrent value. Both retribution and deterrence are irrelevant to AI sentencing. Incapacitation, which *is* relevant for AI sentencing, is not reflected in public service, unless the framework for the specific public service is extremely strict and the offender's delinquent capabilities are incapacitated.

The dominant purpose of public service, however, is rehabilitation through reintegration of the offender into society. Here is the question regarding the applicability of public service: how can public service be imposed on AI systems? To answer this question, we again follow the three-stage model. First, we analyze the roots of the meaning of public service; second, we seek the meaning of these roots for AI systems; and finally, we adjust the punishment to the roots we have revealed for AI systems.

Functionally, public service consists of supervised compensation to society provided by the experience of integration with society. The offender expands his or her experience with society, making integration easier. Broadening the offender's social experience benefits society because it also includes a compensational dimension. Social experience is not exclusive to human offenders. Both corporations and AI systems have strong interactions with the community. Public service may empower and strengthen these interactions, and make them the basis for the required inner change.

For example, a medical expert system, equipped with machine-learning capabilities, is used in a private clinic to diagnose patients. The system was considered negligent, and the court has imposed public service on it. To carry out this penalty, the system may be used by public medical services or public hospitals. This can serve two main goals. First, the system is exposed to more cases, and through machine learning it can refine its functioning. Second, its work can be considered compensation to society for the harm caused by the offense.

At the end of the public service term, the AI system is more experienced, and if the machine-learning process was effective, system performance is improved. Because public service is supervised, regardless of whether it is accompanied by probation, the machine-learning process and other inner processes are directed toward the prevention of further offenses during the public service. By the end of the public service period, the AI system will have contributed time and resources to the benefit of society, and this may be considered compensation for the social harm caused by the commission of the offense. Thus, the public service of AI systems resembles human public service in its substance.

It may be argued that the compensation is in practice contributed by the manufacturers or users of the system, because they are the ones who suffer from the absence of the system's activity. This is true not only for AI systems, but also for human offenders and corporations. When a human offender performs public service, his or her absence is felt by family and friends. When a corporation carries out public service, its resources are not available to its employees, directors, clients, and so on. This absence is part of carrying out the public service, regardless of the identity of the offender, so that AI systems are not unique in this regard.

6.2.5. Fine

Fine is a payment made by the offender to the state treasury. It is not considered compensation because it is not given directly to the victims of the offense, but to society as a whole. If we regard society as the victim of any offense, we may view a fine as a type of general compensation. The fine has evolved from the general remedy of compensation, when the criminal process was still between two individuals (private plaintiff versus defendant).[59] When criminal law became public, the original compensation was converted into a fine. Today, criminal courts may impose both fines and compensations as part of the criminal process.

In the eighteenth century, the fine was not considered a preferred penalty because it was not deemed to be as strong a deterrent as imprisonment.[60] But during the twentieth century, when prisons became overcrowded and the cost of maintaining prisoners in state prisons increased, the penal system was urged to increase the use of fines.[61] To facilitate the efficient collection of fines, in most legal systems the court can impose imprisonment, public service, or confiscation of property in case of non-payment.⁶² The fine imposed is not necessarily proportional to the harm caused by the offense, but to the severity of the offense.

Retribution may be relevant to fines, if the fine is proportional to the social harm and reflects it. Deterrence may also be relevant to fines, if the fine causes a loss of property sufficiently great to deter. But as noted earlier, both retribution and deterrence are irrelevant to AI sentencing. Fines have no prominent rehabilitative value, although paying the fine may require additional work from the offender, allowing less free time to commit offenses. Here is the question regarding the applicability of fines: how can fines be imposed on AI systems? The main difficulty is that AI systems possess no money or other property of their own.

Corporations possess property, so paying fines is the easiest way of imposing a penalty on corporations, but AI systems do not possess property. To answer our question, we again follow the three-stage model. First, we analyze the roots of the meaning of fines; second, we seek the meaning of these roots for AI systems; and finally, we adjust the punishment to the roots we have revealed for AI systems. The result should also include appropriate solutions for ineffective fines, that is, cases in which there are difficulties in collecting the fines.

Functionally, a fine is a forced contribution of valuable property to society. In most cases a fine is expressed as money, but in certain legal systems, other valuable property may be used. In some legal systems, the amount of the fine is determined based on the cost of a work day, week, or month of the defendant, with the fine matching this cost.⁶³ Even if the fine is determined as an absolute sum, in most cases the absence of this sum in the offender's pocket translates into the work hours needed to produce the missing amount. Therefore, a fine generally reflects work hours, days, weeks, or months, depending on the amount and on the offender's wealth or skills.

As we discussed earlier in the context of public service, the productivity of an AI system can also be evaluated in work hours for the community.⁶⁴ It is true that AI systems do not possess property, but they are capable of working, which is valuable and may be assigned a monetary value. For example, the work hour of a medical expert system may be assessed as $500. If a fine of $1,000 is imposed on an AI system, the fine can be translated into two work hours of the system. Therefore, the system can pay by using the only currency it possesses: work hours.

The work hours are contributed to society, in the same way that public service is contributed. When a human offender does not have the required sum of money to pay the fine, other penalties are imposed. One of these is public service, which may be measured by the number of work hours. Using work hours as a measure of payment can serve not only the purpose of paying the fine (by an accurate accounting of the number of hours worked), but also the optional purpose of enforcing the fine together with public service, imprisonment, or any other relevant penalty.

It may be argued that this payment to society is in practice contributed by the manufacturers or users of the AI system, because they are the ones who suffer from the absence of the system's activity while it is working to pay the fine. This is true not only for AI systems, but also for human offenders and corporations. When a human offender pays a fine, the absence of the money (or of the person, if additional work hours are needed to make up for the absence of money) is felt by family and friends. When a corporation pays a fine, its resources are not available to its employees, directors, clients, and so on. This absence is part of paying the fine, regardless of the identity of the offender, so that AI systems are not unique in this regard.

CLOSING THE OPENING EXAMPLE: SENTENCING THE KILLING ROBOT

We opened this chapter with an example that described an AI robot convicted of manslaughter. We can now return to this example and analyze it based on the insights gained from this chapter. After the court has convicted the AI robot for manslaughter, the criminal process continues as if the offender were human, and the next stage is that of sentencing. The prosecution and the defense may argue for a harsh or lenient penalty, and eventually the court decides the appropriate penalty based on the relevant circumstances of the case.

In our example, after assessing the circumstances, the court reaches the conclusion that if the robot were human, the relevant penalty would be three-year prison sentence. If the robot had been a corporation, the court would have imposed three years of imprisonment, which would have been carried out as demonstrated earlier.[65] The same should be the case with the AI robot. Therefore, if the judicial discretion of the court results in the imposition of three years' imprisonment, this is the penalty that must be imposed, regardless of whether the defendant is a human, a corporate, or an AI offender.

After its imposition, the penalty must be carried out. As previously noted, the imposition of imprisonment on AI systems takes the form of depriving them of their liberty to act by restricting their activity for the required term and under strict supervision. During this time, the AI system can be repaired to prevent the commission of further offenses. Repair of the AI system may be more efficient if the system is incapacitated and when it happens under court order. This situation serves both the purposes of rehabilitation and incapacitation, which are the relevant sentencing purposes in the case of AI systems. When the AI system is in custody, restricted, and supervised, its capabilities to offend are incapacitated.

Repairing the AI system through inner changes, initiated by external factors (for example, programmers working under court order) and experienced by the system as a restriction, is substantive rehabilitation because the chances of the system being involved in further delinquency are reduced. The social value of imposing imprisonment on an AI system is real. The dangerous system is being removed from society for repairs, during which time it is not capable of causing further harm to society. When the process is complete, at the end of the term determined by the court, the system can be returned to full activity, the same way that a prisoner who finishes serving a penalty is returned to society.

CONCLUSIONS

Criminal liability for artificial intelligence entities may sound radical. For centuries, criminal liability was considered to be part of an exclusively human universe. The first crack in the concept occurred in the seventeenth century, when corporations were admitted into this exclusive club. Corporate offenders joined human offenders, as both criminal liability and punishments were equally imposed on both. The twentieth century introduced AI technology to humankind. This technology developed rapidly in an attempt to imitate human capabilities. Today, AI systems imitate some human capabilities perfectly; indeed, they outperform humans in many areas. Other human capabilities, however, still cannot be imitated.

Criminal liability does not require that offenders possess all human capabilities, only some. If an AI entity possesses these capabilities, then logically and rationally, criminal liability can be imposed whenever an offense is committed. The idea of criminal liability of AI entities should not be confused with the idea of moral accountability, which is a hugely complex issue not only for machines, but for humans. Morality, in general, has no single definition that is acceptable to all societies and individuals.

Deontological and teleological morality are the most acceptable types, and in many situations they lead in opposite directions, both "moral." The Nazis considered themselves to be deontologically moral, although most other people, societies, and individuals thought otherwise. Because morality is so difficult to assess, moral accountability is not the most appropriate and efficient way of evaluating responsibility in criminal cases. Therefore, society has chosen criminal liability as the most appropriate means for dealing with delinquency. Modern criminal liability is independent from morality of any kind, as well as from the concept of evil. It is imposed in an organized, almost mathematical way.

Criminal liability has definite requirements, not more and not less. If an AI entity meets all these requirements, then there is no reason not to impose criminal liability on it. We have seen that AI entities are capable of meeting these requirements in the same way that human and corporate offenders do (but not animals). We have also seen that all types of criminal liability are relevant to AI technology, and all substantive arguments of criminal liability can be made regarding AI entities. Finally, we have also seen that penalties can be imposed on AI entities in a way that substantively resembles the way in which corporations are being punished.

Society derives the usual benefits of criminal law by imposing criminal liability on AI entities. The conclusion is simple. Either we impose criminal liability on AI entities, or we must change the basic definition of criminal liability as it developed over thousands of years, and abandon the traditional understandings of criminal liability.

NOTES

Preface
1. This is the opening example of chapter 2, analyzed there in detail.

1. The Emergence of *Machina Sapiens Criminalis*
1. See, for example, AAGE GERHARDT DRACHMANN, THE MECHANICAL TECHNOLOGY OF GREEK AND ROMAN ANTIQUITY: A STUDY OF THE LITERARY SOURCES (1963); J. G. LANDELS, ENGINEERING IN THE ANCIENT WORLD (rev. ed. 2000).

2. RENÉ DESCARTES, DISCOURS DE LA MÉTHODE POUR BIEN CONDUIRE SA RAISON ET CHERCHER LA VÉRITÉ DANS LES SCIENCES (1637) (Discourse on the Method of Rightly Conducting One's Reason and of Seeking Truth in the Sciences).

3. Terry Winograd, *Thinking Machines: Can There Be? Are We?*, in THE FOUNDATIONS OF ARTIFICIAL INTELLIGENCE 167, 168 (Derek Pertridge & Yorick Wilks eds., 1990, 2006).

4. THOMAS HOBBES, LEVIATHAN OR THE MATTER, FORME AND POWER OF A COMMON WEALTH ECCLESIASTICALL AND CIVIL III.xxxii.2 (1651): "When a man reasoneth, he does nothing else but conceive a sum total, from addition of parcels; or conceive a remainder . . . These operations are not incident to numbers only, but to all manner of things that can be added together, and taken one out of another . . . the logicians teach the same in consequences of words; adding together two names to make an affirmation, and to affirmations to make a syllogism; and many syllogisms to make a demonstration."

5. GOTTFRIED WILHELM LEIBNIZ, CHARACTERISTICA UNIVERSALIS (1676).

6. N. P. PADHY, ARTIFICIAL INTELLIGENCE AND INTELLIGENT SYSTEMS 4 (2005, 2009).

7. DAN W. PATTERSON, INTRODUCTION TO ARTIFICIAL INTELLIGENCE AND EXPERT SYSTEMS (1990).

8. GEORGE F. LUGER, ARTIFICIAL INTELLIGENCE: STRUCTURES AND STRATEGIES FOR COMPLEX PROBLEM SOLVING (2001).

9. J. R. MCDONALD, G. M. BURT, J. S. ZIELINSKI, & S. D. J. MCARTHUR, INTELLIGENT KNOWLEDGE BASED SYSTEM IN ELECTRICAL POWER ENGINEERING (1997).

10. STUART J. RUSSELL & PETER NORVIG, ARTIFICIAL INTELLIGENCE: A MODERN APPROACH (2002).

11. PATTERSON, *supra* note 7.

12. STEVEN L. TANIMOTO, ELEMENTS OF ARTIFICIAL INTELLIGENCE: AN INTRODUCTION USING LISP (1987).

13. See, for example, Edwina L. Rissland, *Artificial Intelligence and Law: Stepping Stones to a Model of Legal Reasoning*, 99 YALE L.J. 1957, 1961–1964 (1990); ALAN TYREE, EXPERT SYSTEMS IN LAW 7–11 (1989).

14. ROBERT M. GLORIOSO & FERNANDO C. COLON OSORIO, ENGINEERING INTELLIGENT SYSTEMS: CONCEPTS AND APPLICATIONS (1980).

15. PADHY, *supra* note 6, at p. 13.

16. Nick Carbone, South Korea Rolls Out Robotic Prison Guards (Nov. 27, 2011), http://newsfeed.time.com (last visited Jan. 4, 2012); Alex Knapp, South Korean Prison To Feature Robot Guards (Nov. 27, 2011), http://www.forbes.com (last visited Feb. 29, 2012).

17. W. J. Hennigan, *New Drone Has No Pilot Anywhere, So Who's Accountable?*, LOS ANGELES TIMES, Jan. 26, 2012. See also http://www.latimes.com/business (last visited Feb. 29, 2012).

18. EUGENE CHARNIAK & DREW MCDERMOTT, INTRODUCTION TO ARTIFICIAL INTELLIGENCE (1985).

19. GENESIS 3:1–24.

20. RICHARD E. BELLMAN, AN INTRODUCTION TO ARTIFICIAL INTELLIGENCE: CAN COMPUTERS THINK? (1978).

21. JOHN HAUGELAND, ARTIFICIAL INTELLIGENCE: THE VERY IDEA (1985).

22. CHARNIAK & MCDERMOTT, *supra* note 18.

23. ROBERT J. SCHALKOFF, ARTIFICIAL INTELLIGENCE: AN ENGINEERING APPROACH (1990).

24. RAYMOND KURZWEIL, THE AGE OF INTELLIGENT MACHINES (1990).

25. PATRICK HENRY WINSTON, ARTIFICIAL INTELLIGENCE (3d ed. 1992).

26. GEORGE F. LUGER & WILLIAM A. STUBBLEFIELD, ARTIFICIAL INTELLIGENCE: STRUCTURES AND STRATEGIES FOR COMPLEX PROBLEM SOLVING (6th ed. 2008).

27. ELAINE RICH & KEVIN KNIGHT, ARTIFICIAL INTELLIGENCE (2d ed. 1991).

28. PADHY, *supra* note 6, at p. 7.

29. Alan Turing, *Computing Machinery and Intelligence*, 59 MIND 433, 433–460 (1950).

30. Donald Davidson, *Turing's Test, in* MODELLING THE MIND 1 (K. A. Mohyeldin Said, W. H. Newton-Smith, R. Viale, & K. V. Wilkes eds., 1990).

31. Robert M. French, *Subcognition and the Limits of the Turing Test*, 99 MIND 53, 53–54 (1990).

32. MASOUD YAZDANI & AJIT NARAYANAN, ARTIFICIAL INTELLIGENCE: HUMAN EFFECTS (1985).

33. STEVEN L. TANIMOTO, ELEMENTS OF ARTIFICIAL INTELLIGENCE: AN INTRODUCTION USING LISP (1987).

34. JOHN R. SEARLE, MINDS, BRAINS AND SCIENCE 28–41 (1984); John R. Searle, *Minds, Brains & Programs*, 3 BEHAVIORAL & BRAIN SCI. 417 (1980).

35. Roger C. Schank, *What is AI, Anyway?, in* THE FOUNDATIONS OF ARTIFICIAL INTELLIGENCE 3, 4–6 (Derek Pertridge & Yorick Wilks eds., 1990, 2006).

36. DOUGLAS R. HOFSTADTER, GÖDEL, ESCHER, BACH: AN ETERNAL GOLDEN BRAID 539–604 (1979, 1999).

NOTES TO CHAPTER 1 · 181

37. See, for example, DONALD A. WATERMAN, A GUIDE TO EXPERT SYSTEMS (1986): "It wasn't until the late 1970s that AI scientists began to realize something quite important: The problem-solving power of a program comes from the knowledge it possesses, not just from the formalisms and inference schemes it employs. The conceptual breakthrough was made and can be quite simply stated. *To make a program intelligent, provide it with lots of high-quality, specific knowledge about some problem area.*" [original emphasis]; DONALD MICHIE & RORY JOHNSTON, THE CREATIVE COMPUTER (1984); EDWARD A. FEIGENBAUM & PAMELA MCCORDUCK, THE FIFTH GENERATION: ARTIFICIAL INTELLIGENCE AND JAPAN'S COMPUTER CHALLENGE TO THE WORLD (1983).

38. See, for example, the criticism of Winograd, *supra* note 3, at pp. 178–181.

39. Schank, *supra* note 35, pp. 9–12.

40. See PHILLIP N. JOHNSON-LAIRD, MENTAL MODELS 448–477 (1983); But see also COLIN MCGINN, THE PROBLEM OF CONSCIOUSNESS: ESSAYS TOWARDS A RESOLUTION 202, 209–213 (1991).

41. HOWARD GARDNER, THE MIND'S NEW SCIENCE: A HISTORY OF THE COGNITIVE REVOLUTION (1985); MARVIN MINSKY, THE SOCIETY OF MIND (1986); ALLEN NEWELL & HERBERT A. SIMON, HUMAN PROBLEM SOLVING (1972); Winograd, *supra* note 3, at pp. 169–171.

42. MAX WEBER, ECONOMY AND SOCIETY: AN OUTLINE OF INTERPRETIVE SOCIOLOGY (1968); Winograd, *supra* note 3, at pp. 182–183.

43. Daniel C. Dennett, *Evolution, Error, and Intentionality, in* THE FOUNDATIONS OF ARTIFICIAL INTELLIGENCE 190, 190–211 (Derek Pertridge & Yorick Wilks eds., 1990, 2006).

44. Lawrence B. Solum, *Legal Personhood for Artificial Intelligences*, 70 N.C. L. REV. 1231, 1262 (1992); OWEN J. FLANAGAN, JR., THE SCIENCE OF THE MIND 254 (2d ed. 1991); John Haugeland, *Semantic Engines: An Introduction to Mind Design, in* MIND DESIGN 1, 32 (John Haugeland ed., 1981).

45. MONTY NEWBORN, DEEP BLUE (2002).

46. STEPHEN BAKER, FINAL JEOPARDY: MAN VS. MACHINE AND THE QUEST TO KNOW EVERYTHING (2011).

47. VOJISLAV KECMAN, LEARNING AND SOFT COMPUTING, SUPPORT VECTOR MACHINES, NEURAL NETWORKS AND FUZZY LOGIC MODELS (2001).

48. For example, in November 2009, during the Supercomputing Conference in Portland, Oregon (SC09), IBM scientists and others announced that they had succeeded in creating a new algorithm, named "Blue Matter," which possesses the thinking capabilities of a cat. Chris Capps, "Thinking" Supercomputer Now Conscious as a Cat (Nov. 19, 2009), http://www.unexplainable.net (last visited Feb. 29, 2012); International Conference for High Performance Computing, Networking, Storage and Analysis, SC09, http://sc09.supercomputing.org (last visited Feb. 29, 2012). This algorithm collects information from very many units with parallel and distributed connections. The information is integrated, and creates a full image of sensory

information, perception, dynamic action and reaction, and cognition. B. G. FITCH ET AL., BLUE MATTER: AN APPLICATION FRAMEWORK FOR MOLECULAR SIMULATION ON BLUE GENE, IBM Research Report (2003). This platform simulates brain capabilities, and eventually, it is supposed to simulate real thought processes. The final application of this algorithm contains not only analog and digital circuits, metal or plastics, but also protein-based biologic surfaces.

49. See section 1.1.1.

50. TERRY WINOGRAD & FERNANDO C. FLORES, UNDERSTANDING COMPUTERS AND COGNITION: A NEW FOUNDATION FOR DESIGN (1986, 1987); Tom Athanasiou, *High-Tech Politics: The Case of Artificial Intelligence*, 92 SOCIALIST REV. 7, 7–35 (1987); Winograd, *supra* note 3, at p. 181.

51. DAVID LEVY, LOVE AND SEX WITH ROBOTS: THE EVOLUTION OF HUMAN-ROBOT RELATIONSHIPS (2007).

52. Yueh-Hsuan Weng, Chien-Hsun Chen, & Chuen-Tsai Sun, *Toward the Human-Robot Co-Existence Society: On Safety Intelligence for Next Generation Robots*, 1 INT. J. SOC. ROBOT. 267, 267–268 (2009).

53. International Robot Fair 2004 Organizing Office, World Robot Declaration (2004), http://www.prnewswire.co.uk (last visited Feb. 29, 2012).

54. Stefan Lovgren, *A Robot in Every Home by 2020, South Korea Says*, NATIONAL GEOGRAPHIC NEWS (2006), http://news.nationalgeographic.com (last visited Feb. 29, 2012).

55. ARTHUR C. CLARKE, 2001: A SPACE ODYSSEY (1968).

56. 2001: A SPACE ODYSSEY (Metro-Goldwyn-Mayer 1968).

57. TERMINATOR 2: JUDGMENT DAY (TriStar Pictures 1991).

58. THE MATRIX (Warner Bros. Pictures 1999); THE MATRIX RELOADED (Warner Bros. Pictures 2003); THE MATRIX REVOLUTIONS (Warner Bros. Pictures 2003).

59. I, ROBOT (20th Century Fox 2004).

60. Yueh-Hsuan Weng, Chien-Hsun Chen, & Chuen-Tsai Sun, *The Legal Crisis of Next Generation Robots: On Safety Intelligence*, Proceedings of the 11th International Conference on Artificial Intelligence and Law 205–209 (2007).

61. See, for example, Roboethics Roadmap Release 1.1, European Robotics Research Network (2006), http://www.roboethics.org (last visited Feb. 29, 2012).

62. ISAAC ASIMOV, I, ROBOT 40 (1950).

63. Jerry A. Fodor, *Modules, Frames, Fridgeons, Sleeping Dogs, and the Music of the Spheres*, in THE ROBOT'S DILEMMA: THE FRAME PROBLEM IN ARTIFICIAL INTELLIGENCE (Zenon W. Pylyshyn ed., 1987).

64. Isaac Asimov himself wrote in his introduction to *The Rest of Robots* that "[t]here was just enough ambiguity in the Three Laws to provide the conflicts and uncertainties required for new stories, and, to my great relief, it seemed always to be possible to think up a new angle out of the sixty-one words of the Three Laws." ISSAC ASIMOV, THE REST OF ROBOTS 43 (1964).

65. See section 1.1.1.

66. Susan Leigh Anderson, *Asimov's "Three Laws of Robotics" and Machine Metaethics*, 22 AI SOC. 477–493 (2008).

67. Stefan Lovgren, *Robot Codes of Ethics to Prevent Android Abuse, Protect Humans*, NATIONAL GEOGRAPHIC NEWS (2007), http://news.national geographic.com (last visited Feb. 29, 2012).

68. See, for example, GILBERT RYLE, THE CONCEPT OF MIND 54–60 (1954); RICHARD B. BRANDT, ETHICAL THEORY 389 (1959); JOHN J. C. SMART & BERNARD WILLIAMS, UTILITARIANISM — FOR AND AGAINST 18–20 (1973).

69. See, for example, Justine Miller, *Criminal Law—An Agency for Social Control*, 43 YALE L.J. 691 (1934); Jack P. Gibbs, *A Very Short Step toward a General Theory of Social Control* AM. B. FOUND RES. J. 607 (1985); K. W. Lidstone, *Social Control and the Criminal Law*, 27 BRIT. J. CRIMINOLOGY 31 (1987); Justice Ellis, *Criminal Law as an Instrument of Social Control*, 17 VICTORIA U. WELLINGTON L. REV. 319 (1987).

70. See section 1.1.2.

71. For the concept of turning disadvantages into advantages or *advantaging the disadvantages* in industrial and private use, see section 1.1.3.

72. HANS MORAVEC, ROBOT: MERE MACHINE TO TRANSCENDENT MIND (1999).

73. DAVID LEVY, ROBOTS UNLIMITED: LIFE IN A VIRTUAL AGE (2006).

74. For wider aspect, see, for example, Mary Anne Warren, *On the Moral and Legal Status of Abortion*, *in* ETHICS IN PRACTICE (Hugh Lafollette ed., 1997); IMMANUEL KANT, OUR DUTIES TO ANIMALS (1780).

75. David Lyons, *Open Texture and the Possibility of Legal Interpretation*, 18 LAW PHIL. 297, 297–309 (1999).

76. See, for example, EXODUS 22:1, 4, 9, 10; 23:4, 12.

77. EXODUS 21:28–32.

78. Jonathan M. E. Gabbai, Complexity and the Aerospace Industry: Understanding Emergence by Relating Structure to Performance Using Multi-Agent Systems (PhD Thesis, University of Manchester, 2005).

79. Craig W. Reynolds, *Herds and Schools: A Distributed Behavioral Model*, 21 COMPUT. GRAPH. 25–34 (1987).

80. Wyatt S. Newman, *Automatic Obstacle Avoidance at High Speeds via Reflex Control*, Proceedings of the 1989 IEEE International Conference on Robotics and Automation 1104 (1989).

81. See, for example, Gabriel Hallevy, *Unmanned Vehicles—Subordination to Criminal Law under the Modern Concept of Criminal Liability*, 21 J. L. INF. & SCI. 311 (2011).

82. See, for example, State v. Stewart, 624 N.W.2d 585 (Minn.2001); Wheatley v. Commonwealth, 26 Ky.L.Rep. 436, 81 S.W. 687 (1904); State v. Follin, 263 Kan. 28, 947 P.2d 8 (1997).

83. Andrew G. Brooks & Ronald C. Arkin, *Behavioral Overlays for Non-Verbal Communication Expression on a Humanoid Robot*, 22 AUTON. ROBOTS 55, 55–74 (2007).

84. LAWRENCE LESSIG, CODE AND OTHER LAWS OF CYBERSPACE (1999).

85. See, for example, Steven J. Wolhandler, *Voluntary Active Euthanasia for the Terminally Ill and the Constitutional Right to Privacy*, 69 CORNELL L. REV. 363 (1984); Harold L. Hirsh & Richard E. Donovan, *The Right to Die: Medico-Legal Implications of In re Quinlan*, 30 RUTGERS L. REV. 267 (1977); Susan M. Allan, *No Code Orders v. Resuscitation: The Decision to Withhold Life-Prolonging Treatment from the Terminally Ill*, 26 WAYNE L. REV. 139 (1980).

86. For the structure of the principle of legality in criminal law, see GABRIEL HALLEVY, A MODERN TREATISE ON THE PRINCIPLE OF LEGALITY IN CRIMINAL LAW 5–8 (2010).

87. *Id.* at pp. 20–46.

88. *Id.* at pp. 67–78.

89. See, for example, in *Transcript of Proceedings of Nuremberg Trials*, 41 AM. J. INT'L L. 1–16 (1947).

90. HALLEVY, *supra* note 86, at pp. 97–118.

91. *Id.* at pp. 118–129.

92. *Id.* at pp. 135–137.

93. *Id.* at pp. 137–138.

94. *Id.* at pp. 138–141.

95. See, for example, sub-article 58(c)(1) of the Soviet Penal Code of 1926 as amended in 1950. This sub-article provided that mature relatives of the first degree of convicted traitor are punished with five years of exile.

96. See *supra* note 89.

97. Scales v. United States, 367 U.S. 203, 81 S.Ct. 1469, 6 L.Ed.2d 782 (1961); Larsonneur, (1933) 24 Cr. App. R. 74, 97 J.P. 206, 149 L.T. 542; ANDREW ASHWORTH, PRINCIPLES OF CRIMINAL LAW 106–107 (5th ed. 2006); Anderson v. State, 66 Okl.Cr. 291, 91 P.2d 794 (1939); State v. Asher, 50 Ark. 427, 8 S.W. 177 (1888); Peebles v. State, 101 Ga. 585, 28 S.E. 920 (1897); Howard v. State, 73 Ga.App. 265, 36 S.E.2d 161 (1945); Childs v. State, 109 Nev. 1050, 864 P.2d 277 (1993).

98. See, for example, Smith v. State, 83 Ala. 26, 3 So. 551 (1888); People v. Brubaker, 53 Cal.2d 37, 346 P.2d 8 (1959); State v. Barker, 128 W.Va. 744, 38 S.E.2d 346 (1946).

99. See, for example, Commonwealth v. Herd, 413 Mass. 834, 604 N.E.2d 1294 (1992); State v. Curry, 45 Ohio St.3d 109, 543 N.E.2d 1228 (1989); State v. Barrett, 768 A.2d 929 (R.I.2001); State v. Lockhart, 208 W.Va. 622, 542 S.E.2d 443 (2000).

100. See, for example, Beason v. State, 96 Miss. 165, 50 So. 488 (1909); State v. Nickelson, 45 La.Ann. 1172, 14 So. 134 (1893); Commonwealth v. Mead, 92 Mass. 398 (1865); Willet v. Commonwealth, 76 Ky. 230 (1877); Scott v. State, 71 Tex.Crim.R. 41, 158 S.W. 814 (1913); Price v. State, 50 Tex. Crim.R. 71, 94 S.W. 901 (1906).

101. See, for example, Elk v. United States, 177 U.S. 529, 20 S.Ct. 729, 44

L.Ed. 874 (1900); State v. Bowen, 118 Kan. 31, 234 P. 46 (1925); Hughes v. Commonwealth, 19 Ky.L.R. 497, 41 S.W. 294 (1897); People v. Cherry, 307 N.Y. 308, 121 N.E.2d 238 (1954); State v. Hooker, 17 Vt. 658 (1845); Commonwealth v. French, 531 Pa. 42, 611 A.2d 175 (1992).

102. GABRIEL HALLEVY, THE MATRIX OF DERIVATIVE CRIMINAL LIABILITY 18–24 (2012).

103. Dugdale, (1853) 1 El. & Bl. 435, 118 E.R. 499, 500: "the mere intent cannot constitute a misdemeanour when unaccompanied with any act"; *Ex parte* Smith, 135 Mo. 223, 36 S.W. 628 (1896); Proctor v. State, 15 Okl.Cr. 338, 176 P. 771 (1918); State v. Labato, 7 N.J. 137, 80 A.2d 617 (1951); Lambert v. State, 374 P.2d 783 (Okla.Crim.App.1962); *In re* Leroy, 285 Md. 508, 403 A.2d 1226 (1979).

104. For the principle of legality in criminal law, see HALLEVY, *supra* note 86.

105. *Id.* at pp. 135–137.

106. *Id.* at pp. 49–80.

107. *Id.* at pp. 81–132.

108. See, for example, in the United States, Robinson v. California, 370 U.S. 660, 82 S.Ct. 1417, 8 L.Ed.2d 758 (1962).

109. GLANVILLE WILLIAMS, CRIMINAL LAW: THE GENERAL PART sec. 11 (2nd ed. 1961).

110. ANDREW ASHWORTH, PRINCIPLES OF CRIMINAL LAW 157–158, 202 (5th ed. 2006).

111. G. R. Sullivan, *Knowledge, Belief, and Culpability*, in CRIMINAL LAW THEORY — DOCTRINES OF THE GENERAL PART 207, 214 (Stephen Shute & A. P. Simester eds., 2005).

112. See, for example, article 2.02(5) of The American Law Institute, Model Penal Code—Official Draft and Explanatory Notes 22 (1962, 1985), which provides, "When the law provides that negligence suffices to establish an element of an offense, such element also is established if a person acts purposely, knowingly or recklessly. When recklessness suffices to establish an element, such element also is established if a person acts purposely or knowingly. When acting knowingly suffices to establish an element, such element also is established if a person acts purposely."

113. William S. Laufer, *Corporate Bodies and Guilty Minds*, 43 EMORY L.J. 647 (1994); Kathleen F. Brickey, *Corporate Criminal Accountability: A Brief History and an Observation*, 60 WASH. U. L. Q. 393 (1983).

114. WILLIAM SEARLE HOLDSWORTH, A HISTORY OF ENGLISH LAW 475–476 (1923).

115. William Searle Holdsworth, *English Corporation Law in the 16th and 17th Centuries*, 31 YALE L.J. 382 (1922).

116. WILLIAM ROBERT SCOTT, THE CONSTITUTION AND FINANCE OF ENGLISH, SCOTTISH AND IRISH JOINT-STOCK COMPANIES TO 1720 462 (1912).

117. BISHOP CARLETON HUNT, THE DEVELOPMENT OF THE BUSINESS CORPORATION IN ENGLAND 1800–1867 6 (1963).

118. See, for example, 6 Geo. I, c.18 (1719).

119. New York & G.L.R. Co. v. State, 50 N.J.L. 303, 13 A. 1 (1888); People v. Clark, 8 N.Y.Cr. 169, 14 N.Y.S. 642 (1891); State v. Great Works Mill. & Mfg. Co., 20 Me. 41, 37 Am.Dec.38 (1841); Commonwealth v. Proprietors of New Bedford Bridge, 68 Mass. 339 (1854); Commonwealth v. New York Cent. & H. River R. Co., 206 Mass. 417, 92 N.E. 766 (1910).

120. John C. Coffee, Jr., *"No Soul to Damn: No Body to Kick": An Unscandalised Inquiry into the Problem of Corporate Punishment*, 79 MICH. L. REV. 386 (1981).

121. Langforth Bridge, (1635) Cro. Car. 365, 79 E.R. 919.

122. Clifton (Inhabitants), (1794) 5 T.R. 498, 101 E.R. 280; Great Broughton (Inhabitants), (1771) 5 Burr. 2700, 98 E.R. 418; Stratford-upon-Avon Corporation, (1811) 14 East 348, 104 E.R. 636; Liverpool (Mayor), (1802) 3 East 82, 102 E.R. 529; Saintiff, (1705) 6 Mod. 255, 87 E.R. 1002.

123. Severn and Wye Railway Co., (1819) 2 B. & Ald. 646, 106 E.R. 501; Birmingham, &c., Railway Co., (1842) 3 Q. B. 223, 114 E.R. 492; New York Cent. & H.R.R. v. United States, 212 U.S. 481, 29 S.Ct. 304, 53 L.Ed. 613 (1909); United States v. Thompson-Powell Drilling Co., 196 F.Supp. 571 (N.D.Tex.1961); United States v. Dye Construction Co., 510 F.2d 78 (10th Cir.1975); United States v. Carter, 311 F.2d 934 (6th Cir.1963); State v. I. & M. Amusements, Inc., 10 Ohio App.2d 153, 226 N.E.2d 567 (1966).

124. United States v. Alaska Packers' Association, 1 Alaska 217 (1901).

125. United States v. John Kelso Co., 86 F. 304 (Cal.1898); Lennard's Carrying Co. Ltd. v. Asiatic Petroleum Co. Ltd., [1915] A.C. 705.

126. Director of Public Prosecutions v. Kent and Sussex Contractors Ltd., [1944] K.B. 146, [1944] 1 All E.R. 119; I.C.R. Haulage Ltd., [1944] K.B. 551, [1944] 1 All E.R. 691; Seaboard Offshore Ltd. v. Secretary of State for Transport, [1994] 2 All E.R. 99, [1994] 1 W.L.R. 541, [1994] 1 Lloyd's Rep. 593.

127. Granite Construction Co. v. Superior Court, 149 Cal.App.3d 465, 197 Cal.Rptr. 3 (1983); Commonwealth v. Fortner L.P. Gas Co., 610 S.W.2d 941 (Ky.App.1980); Commonwealth v. McIlwain School Bus Lines, Inc., 283 Pa.Super. 1, 423 A.2d 413 (1980); Gerhard O. W. Mueller, *Mens Rea and the Corporation—A Study of the Model Penal Code Position on Corporate Criminal Liability*, 19 U. PITT. L. REV. 21 (1957).

128. Hartson v. People, 125 Colo. 1, 240 P.2d 907 (1951); State v. Pincus, 41 N.J.Super. 454, 125 A.2d 420 (1956); People v. Sakow, 45 N.Y.2d 131, 408 N.Y.S.2d 27, 379 N.E.2d 1157 (1978).

129. See section 6.1.2.

2. AI Criminal Liability for Intentional Offenses

1. The facts of this event are based on the overview in Yueh-Hsuan Weng, Chien-Hsun Chen, & Chuen-Tsai Sun, *Toward the Human-Robot Co-Existence Society: On Safety Intelligence for Next Generation Robots*, 1 INT. J. SOC. ROBOT 267, 273 (2009).

2. See section 1.3.2.

3. Fain v. Commonwealth, 78 Ky. 183, 39 Am.Rep. 213 (1879); Tift v. State, 17 Ga.App. 663, 88 S.E. 41 (1916); People v. Decina, 2 N.Y.2d 133, 157 N.Y.S.2d 558, 138 N.E.2d 799 (1956); Mason v. State, 603 P.2d 1146 (Okl. Crim.App.1979); State v. Burrell, 135 N.H. 715, 609 A.2d 751 (1992); Bonder v. State, 752 A.2d 1169 (Del.2000).

4. The same is true with some other definitions, of which the most popular is "willed muscular movement." See HERBERT L. A. HART, PUNISHMENT AND RESPONSIBILITY: ESSAYS IN THE PHILOSOPHY OF LAW 101 (1968); OLIVER W. HOLMES, THE COMMON LAW 54 (1881, 1923); ANTONY ROBIN DUFF, CRIMINAL ATTEMPTS 239–263 (1996); JOHN AUSTIN, THE PROVINCE OF JURISPRUDENCE DETERMINED (1832, 2000); GUSTAV RADBRUCH, DER HANDLUNGSBEGRIFF IN SEINER BEDEUTUNG FÜR DAS STRAFRECHTSSYSTEM 75, 98 (1904); CLAUS ROXIN, STRAFRECHT—ALLGEMEINER TEIL I 239–255 (4 Auf. 2006); BGH 3, 287.

5. See, for example, Bolden v. State, 171 S.W.3d 785 (2005); United States v. Meyers, 906 F. Supp. 1494 (1995); United States v. Quaintance, 471 F. Supp.2d 1153 (2006).

6. Scott T. Noth, *A Penny for Your Thoughts: Post-Mitchell Hate Crime Laws Confirm a Mutating Effect upon Our First Amendment and the Government's Role in Our Lives*, 10 REGENT U. L. REV. 167 (1998); HENRY HOLT, TELEKINESIS (2005); PAMELA RAE HEATH, THE PK ZONE: A CROSS-CULTURAL REVIEW OF PSYCHOKINESIS (PK) (2003).

7. JOHN AUSTIN, THE PROVINCE OF JURISPRUDENCE DETERMINED (1832, 2000).

8. See, for example, People v. Heitzman, 9 Cal.4th 189, 37 Cal.Rptr.2d 236, 886 P.2d 1229 (1994); State v. Wilson, 267 Kan. 550, 987 P.2d 1060 (1999).

9. Rollin M. Perkins, *Negative Acts in Criminal Law*, 22 IOWA L. REV. 659 (1937); Graham Hughes, *Criminal Omissions*, 67 YALE L.J. 590 (1958); Lionel H. Frankel, *Criminal Omissions: A Legal Microcosm*, 11 WAYNE L. REV. 367 (1965).

10. P. R. Glazebrook, *Criminal Omissions: The Duty Requirement in Offences Against the Person*, 55 L. Q. REV. 386 (1960); Andrew Ashworth, *The Scope of Criminal Liability for Omissions*, 84 L. Q. REV. 424, 441 (1989).

11. Lane v. Commonwealth, 956 S.W.2d 874 (Ky.1997); State v. Jackson, 137 Wash.2d 712, 976 P.2d 1229 (1999); Rachel S. Zahniser, *Morally and Legally: A Parent's Duty to Prevent the Abuse of a Child as Defined by* Lane v. Commonwealth, 86 KY. L.J. 1209 (1998).

12. Mavji, [1987] 2 All E.R. 758, [1987] 1 W.L.R. 1388, [1986] S.T.C. 508, Cr. App. Rep. 31, [1987] Crim. L.R. 39; Firth, (1990) 91 Cr. App. Rep. 217, 154 J.P. 576, [1990] Crim. L.R. 326.

13. See, for example, section 2.01(3) of The American Law Institute, Model Penal Code—Official Draft and Explanatory Notes (1962, 1985).

14. GABRIEL HALLEVY, THE MATRIX OF DERIVATIVE CRIMINAL LIABILITY 171–184 (2012).

15. See, for example, Pierson v. State, 956 P.2d 1119 (Wyo.1998).

16. See, for example, State v. Dubina, 164 Conn. 95, 318 A.2d 95 (1972); State v. Bono, 128 N.J.Super. 254, 319 A.2d 762 (1974); State v. Fletcher, 322 N.C. 415, 368 S.E.2d 633 (1988).

17. S. Z. Feller, *Les Délits de Mise en Danger*, 40 REV. INT. DE DROIT PÉNAL 179 (1969).

18. This is the results component of all homicide offenses. See SIR EDWARD COKE, INSTITUTIONS OF THE LAWS OF ENGLAND — THIRD PART 47 (6th ed. 1681, 1817, 2001): "Murder is when a man of sound memory, and of the age of discretion, unlawfully killeth within any county of the realm any reasonable creature in rerum natura under the king's peace, with malice aforethought, either expressed by the party or implied by law, [so as the party wounded, or hurt, etc die of the wound or hurt, etc within a year and a day after the same]."

19. Henderson v. Kibbe, 431 U.S. 145, 97 S.Ct. 1730, 52 L.Ed.2d 203 (1977); Commonwealth v. Green, 477 Pa. 170, 383 A.2d 877 (1978); State v. Crocker, 431 A.2d 1323 (Me.1981); State v. Martin, 119 N.J. 2, 573 A.2d 1359 (1990).

20. See, for example, Wilson v. State, 24 S.W. 409 (Tex.Crim.App.1893); Henderson v. State, 11 Ala.App. 37, 65 So. 721 (1914); Cox v. State, 305 Ark. 244, 808 S.W.2d 306 (1991); People v. Bailey, 451 Mich. 657, 549 N.W.2d 325 (1996).

21. Morton J. Horwitz, *The Rise and Early Progressive Critique of Objective Causation*, in THE POLITICS OF LAW: A PROGRESSIVE CRITIQUE 471 (David Kairys ed., 3d ed. 1998); Benge, (1865) 4 F. & F. 504, 176 E.R. 665; Longbottom, (1849) 3 Cox C. C. 439.

22. Jane Stapelton, *Law, Causation and Common Sense*, 8 OXFORD J. LEGAL STUD. 111 (1988).

23. JEROME HALL, GENERAL PRINCIPLES OF CRIMINAL LAW 70–77 (2d ed. 1960, 2005); DAVID ORMEROD, SMITH & HOGAN CRIMINAL LAW 91–92 (11th ed. 2005); G., [2003] U.K.H.L. 50, [2004] 1 A.C. 1034, [2003] 3 W.L.R. 1060, [2003] 4 All E.R. 765, [2004] 1 Cr. App. Rep. 21, (2003) 167 J.P. 621, [2004] Crim. L. R. 369.

24. Sweet v. Parsley, [1970] A.C. 132, [1969] 1 All E.R. 347, [1969] 2 W.L.R. 470, 133 J.P. 188, 53 Cr. App. Rep. 221, 209 E.G. 703, [1969] E.G.D. 123.

25. See sections 1.3.2 and 2.1.

26. See, for example, *supra* note 16.

27. For the definition of *circumstances*, see section 2.1.2.

28. For the definition of *results*, see section 2.1.3.

29. This component of awareness functions also as the legal causal connection in *mens rea* offenses, but this function has no additional significance in this context.

30. *Specific intent* is sometimes mistakenly referred to as *intent* in order to differentiate it from *general intent*, which is normally used to express *mens rea*.

31. See section 5.2.2.

32. SIR GERALD GORDON, THE CRIMINAL LAW OF SCOTLAND 61 (1st ed. 1967); Treacy v. Director of Public Prosecutions, [1971] A.C. 537, 559, [1971] 1 All E.R. 110, [1971] 2 W.L.R. 112, 55 Cr. App. Rep. 113, 135 J.P. 112.

33. William G. Lycan, *Introduction*, *in* MIND AND COGNITION 3, 3–13 (William G. Lycan ed., 1990).

34. See, for example, WILLIAM JAMES, THE PRINCIPLES OF PSYCHOLOGY (1890).

35. See, for example, BERNARD BAARS, IN THE THEATRE OF CONSCIOUSNESS (1997).

36. HERMANN VON HELMHOLTZ, THE FACTS OF PERCEPTION (1878).

37. United States v. Youts, 229 F.3d 1312 (10th Cir.2000); State v. Sargent, 156 Vt. 463, 594 A.2d 401 (1991); United States v. Spinney, 65 F.3d 231 (1st Cir.1995); State v. Wyatt, 198 W.Va. 530, 482 S.E.2d 147 (1996); United States v. Wert-Ruiz, 228 F.3d 250 (3rd Cir.2000); Rollin M. Perkins, *"Knowledge" as a Mens Rea Requirement*, 29 HASTINGS L.J. 953, 964 (1978).

38. United States v. Jewell, 532 F.2d 697 (9th Cir.1976); United States v. Ladish Malting Co., 135 F.3d 484 (7th Cir.1998).

39. Paul Weiss, *On the Impossibility of Artificial Intelligence*, 44 REV. METAPHYSICS 335, 340 (1990).

40. See, for example, TIM MORRIS, COMPUTER VISION AND IMAGE PROCESSING (2004); MILAN SONKA, VACLAV HLAVAC, & ROGER BOYLE, IMAGE PROCESSING, ANALYSIS, AND MACHINE VISION (2008).

41. WALTER W. SOROKA, ANALOG METHODS IN COMPUTATION AND SIMULATION (1954).

42. See, for example, DAVID MANNERS & TSUGIO MAKIMOTO, LIVING WITH THE CHIP (1995).

43. For robot guards, see section 1.1.1.

44. For the endlessness of the quest for *machina sapiens*, see section 1.1.2.

45. *In re* Winship, 397 U.S. 358, 90 S.Ct. 1068, 25 L.Ed.2d 368 (1970).

46. United States v. Heredia, 483 F.3d 913 (2006); United States v. Ramon-Rodriguez, 492 F.3d 930 (2007); Saik, [2006] U.K.H.L. 18, [2007] 1 A.C. 18; Da Silva, [2006] E.W.C.A. Crim. 1654 , [2006] 4 All E.R. 900, [2006] 2 Cr. App. Rep. 517; Evans v. Bartlam, [1937] A.C. 473, 479, [1937] 2 All E.R. 646; G. R. Sullivan, *Knowledge, Belief, and Culpability*, *in* CRIMINAL LAW THEORY — DOCTRINES OF THE GENERAL PART 207, 213–214 (Stephen Shute & A. P. Simester eds., 2005).

47. State v. Pereira, 72 Conn. App. 545, 805 A.2d 787 (2002); Thompson v. United States, 348 F.Supp.2d 398 (2005); Virgin Islands v. Joyce, 210 F. App. 208 (2006).

48. See, for example, United States v. Doe, 136 F.3d 631 (9th Cir.1998); State v. Audette, 149 Vt. 218, 543 A.2d 1315 (1988); Ricketts v. State, 291 Md. 701, 436 A.2d 906 (1981); State v. Rocker, 52 Haw. 336, 475 P.2d 684 (1970); State v. Hobbs, 252 Iowa 432, 107 N.W.2d 238 (1961); State v. Daniels, 236 La. 998, 109 So.2d 896 (1958).

49. People v. Disimone, 251 Mich.App. 605, 650 N.W.2d 436 (2002); Carter v. United States, 530 U.S. 255, 120 S.Ct. 2159, 147 L.Ed.2d 203 (2000); State v. Neuzil, 589 N.W.2d 708 (Iowa 1999); People v. Henry, 239 Mich.App. 140, 607 N.W.2d 767 (1999); Frey v. United States, 708 So.2d 918 (Fla.1998); United States v. Randolph, 93 F.3d 656 (9th Cir.1996); United States v. Torres, 977 F.2d 321 (7th Cir.1992).

50. Schmidt v. United States, 133 F. 257 (9th Cir.1904); State v. Ehlers, 98 N.J.L. 263, 119 A. 15 (1922); United States v. Pomponio, 429 U.S. 10, 97 S.Ct. 22, 50 L.Ed.2d 12 (1976); State v. Gray, 221 Conn. 713, 607 A.2d 391 (1992); State v. Mendoza, 709 A.2d 1030 (R.I.1998).

51. LUDWIG WITTGENSTEIN, PHILOSOPHISCHE UNTERSUCHUNGEN §§629–660 (1953).

52. State v. Ayer, 136 N.H. 191, 612 A.2d 923 (1992); State v. Smith, 170 Wis.2d 701, 490 N.W.2d 40 (App.1992).

53. Studstill v. State, 7 Ga. 2 (1849); Glanville Williams, *Oblique Intention*, 46 CAMB. L.J. 417 (1987).

54. People v. Smith, 57 Cal. App. 4th 1470, 67 Cal. Rptr. 2d 604 (1997); Wieland v. State, 101 Md. App. 1, 643 A.2d 446 (1994).

55. Stephen Shute, *Knowledge and Belief in the Criminal Law*, in CRIMINAL LAW THEORY — DOCTRINES OF THE GENERAL PART 182–187 (Stephen Shute & A. P. Simester eds., 2005); ANTONY KENNY, WILL, FREEDOM AND POWER 42–43 (1975); JOHN R. SEARLE, THE REDISCOVERY OF MIND 62 (1992); ANTONY KENNY, WHAT IS FAITH? 30–31 (1992); State v. VanTreese, 198 Iowa 984, 200 N.W. 570 (1924), but see Montgomery v. Commonwealth, 189 Ky. 306, 224 S.W. 878 (1920); State v. Murphy, 674 P.2d 1220 (Utah.1983); State v. Blakely, 399 N.W.2d 317 (S.D.1987).

56. See, for example, Ned Block, *What Intuitions About Homunculi Don't Show*, 3 BEHAVIORAL & BRAIN SCI. 425 (1980); Bruce Bridgeman, *Brains + Programs = Minds*, 3 BEHAVIORAL & BRAIN SCI. 427 (1980).

57. See, for example, FENG-HSIUNG HSU, BEHIND DEEP BLUE: BUILDING THE COMPUTER THAT DEFEATED THE WORLD CHESS CHAMPION (2002); DAVID LEVY & MONTY NEWBORN, HOW COMPUTERS PLAY CHESS (1991).

58. See section 2.2.2.

59. For the endlessness of the quest for *machina sapiens*, see section 1.1.2.

60. See section 1.3.2.

61. Russian roulette is a game of chance in which participants place a single round in a gun, spin the cylinder, place the muzzle against their own head, and pull the trigger.

62. G., [2003] U.K.H.L. 50, [2003] 4 All E.R. 765, [2004] 1 Cr. App. Rep. 237, 167 J.P. 621, [2004] Crim. L.R. 369, [2004] 1 A.C. 1034; Victor Tadors, *Recklessness and the Duty to Take Care*, in CRIMINAL LAW THEORY — DOCTRINES OF THE GENERAL PART 227 (Stephen Shute & A. P. Simester eds., 2005); Gardiner, [1994] Crim. L.R. 455.

63. VOJISLAV KECMAN, LEARNING AND SOFT COMPUTING, SUPPORT VECTOR MACHINES, NEURAL NETWORKS AND FUZZY LOGIC MODELS (2001).

64. K., [2001] U.K.H.L. 41, [2002] 1 A.C. 462; B. v. Director of Public Prosecutions, [2000] 2 A.C. 428, [2000] 1 All E.R. 833, [2000] 2 W.L.R. 452, [2000] 2 Cr. App. Rep. 65, [2000] Crim. L.R. 403.

65. Robert N. Shapiro, *Of Robots, Persons, and the Protection of Religious Beliefs*, 56 S. CAL. L. R. 1277, 1286–1290 (1983).

66. Nancy Sherman, *The Place of the Emotions in Kantian Morality*, in IDENTITY, CHARACTER, AND MORALITY 145, 149–162 (Owen Flanagan & Amelie O. Rotry eds., 1990); Aaron Sloman, Motives, *Mechanisms, and Emotions, in* THE PHILOSOPHY OF ARTIFICIAL INTELLIGENCE 231, 231–232 (Margaret A. Boden ed., 1990).

67. JOHN FINNIS, NATURAL LAW AND NATURAL RIGHTS 85–90 (1980).

68. Lawrence B. Solum, *Legal Personhood for Artificial Intelligences*, 70 N.C. L. REV. 1231, 1262 (1992).

69. OWEN J. FLANAGAN, JR., THE SCIENCE OF THE MIND 224–241 (2d ed. 1991).

70. See, for example, DAVID LYONS, FORMS AND LIMITS OF UTILITARIANISM (1965).

71. See Solum, *supra* note 68, at pp. 1258–1262.

72. See section 1.3.3.

73. EDWARD O. WILSON, SOCIOBIOLOGY: THE NEW SYNTHESIS 120 (1975).

74. ROGER J. SULLIVAN, IMMANUEL KANT'S MORAL THEORY 68 (1989).

75. RAY JACKENDOFF, CONSCIOUSNESS AND THE COMPUTATIONAL MIND 275–327 (1987); COLIN MCGINN, THE PROBLEM OF CONSCIOUSNESS: ESSAYS TOWARDS A RESOLUTION 202–213 (1991).

76. DANIEL C. DENNETT, BRAINSTORMS 149–150 (1978).

77. See section 1.3.2.

78. DANIEL C. DENNETT, THE INTENTIONAL STANCE 327–328 (1987).

79. See, for example, People v. Marshall, 362 Mich. 170, 106 N.W.2d 842 (1961); State v. Gartland, 304 Mo. 87, 263 S.W. 165 (1924); State v. Etzweiler, 125 N.H. 57, 480 A.2d 870 (1984); People v. Kemp, 150 Cal.App.2d 654, 310 P.2d 680 (1957); State v. Hopkins, 147 Wash. 198, 265 P. 481 (1928); State v. Foster, 202 Conn. 520, 522 A.2d 277 (1987); State v. Garza, 259 Kan. 826, 916 P.2d 9 (1996); Mendez v. State, 575 S.W.2d 36 (Tex.Crim.App.1979).

80. Francis Bowes Sayre, *Criminal Responsibility for the Acts of Another*, 43 HARV. L. REV. 689, 689–690 (1930).

81. DIGESTA, 9.4.2; ULPIAN, 18 ad ed.; OLIVIA F. ROBINSON, THE CRIMINAL LAW OF ANCIENT ROME 15–16 (1995).

82. Y.BB. 32–33 Edw. I (R. S.), 318, 320 (1304); Seaman v. Browning, (1589) 4 Leonard 123, 74 E.R. 771.

83. 13 Edw. I, St. I, c.2, art. 3, c.II, c.43 (1285). See also FREDERICK POLLOCK & FREDERICK WILLIAM MAITLAND, THE HISTORY OF ENGLISH LAW BEFORE THE TIME OF EDWARD I 533 (rev. 2d ed. 1898); Oliver W. Holmes, *Agency*, 4 HARV. L. REV. 345, 356 (1891).

84. Kingston v. Booth, (1685) Skinner 228, 90 E.R. 105.

85. Boson v. Sandford, (1690) 2 Salkeld 440, 91 E.R. 382: "The owners are liable in respect of the freight, and as employing the master; for whoever employs another is answerable for him, and undertakes for his care to all that make use of him"; Turberwill v. Stamp, (1697) Skinner 681, 90 E.R. 303; Middleton v. Fowler, (1699) 1 Salkeld 282, 91 E.R. 247; Jones v. Hart, (1699) 2 Salkeld 441, 91 E.R. 382; Hern v. Nichols, (1708) 1 Salkeld 289, 91 E.R. 256.

86. Sayre, *supra* note 80, at pp. 693–694; WILLIAM PALEY, A TREATISE ON THE LAW OF PRINCIPAL AND AGENT (2d ed. 1847); Huggins, (1730) 2 Strange 882, 93 E.R. 915; Holbrook, (1878) 4 Q.B.D. 42; Chisholm v. Doulton, (1889) 22 Q.B.D. 736; Hardcastle v. Bielby, [1892] 1 Q.B. 709.

87. Glanville Williams, *Innocent Agency and Causation*, 3 CRIM. L. F. 289 (1992); Peter Alldridge, *The Doctrine of Innocent Agency*, 2 CRIM. L. F. 45 (1990).

88. State v. Silva-Baltazar, 125 Wash.2d 472, 886 P.2d 138 (1994); GLANVILLE WILLIAMS, CRIMINAL LAW: THE GENERAL PART 395 (2d ed. 1961).

89. Maxey v. United States, 30 App. D.C. 63, 80 (App. D.C. 1907).

90. Johnson v. State, 142 Ala. 70, 71 (1904).

91. United States v. Bryan, 483 F.2d 88, 92 (3d Cir. 1973).

92. Commonwealth v. Hill, 11 Mass. 136 (1814); Michael, (1840) 2 Mood. 120, 169 E.R. 48.

93. Johnson v. State, 38 So. 182, 183 (Ala. 1904); People v. Monks, 24 P.2d 508, 511 (Cal. Dist. Ct. App. 1933).

94. United States v. Bryan, 483 F.2d 88, 92 (3d Cir. 1973); Boushea v. United States, 173 F.2d 131, 134 (8th Cir. 1949); People v. Mutchler, 140 N.E. 820, 823 (Ill. 1923); State v. Runkles, 605 A.2d 111, 121 (Md. 1992); Parnell v. State, 912 S.W.2d 422, 424 (Ark. 1996); State v. Thomas, 619 S.W.2d 513, 514 (Tenn. 1981).

95. Dusenbery v. Commonwealth, 263 S.E.2d 392 (Va. 1980).

96. United States v. Tobon-Builes, 706 F.2d 1092, 1101 (11th Cir. 1983); United States v. Ruffin, 613 F.2d 408, 411 (2d Cir. 1979).

97. See Solum, *supra* note 68, at p. 1237.

98. The AI entity is used as an instrument and not as a participant, although it uses its features of processing information. See, for example, George R. Cross & Cary G. Debessonet, *An Artificial Intelligence Application in the Law: CCLIPS, A Computer Program that Processes Legal Information*, 1 HIGH TECH. L.J. 329 (1986).

99. Andrew J. Wu, *From Video Games to Artificial Intelligence: Assigning Copyright Ownership to Works Generated by Increasingly Sophisticated Computer Programs*, 25 AIPLA Q. J. 131 (1997); Timothy L. Butler, *Can a Computer Be an Author—Copyright Aspects of Artificial Intelligence*, 4 COMM. ENT. L. S. 707 (1982).

100. NICOLA LACEY & CELIA WELLS, RECONSTRUCTING CRIMINAL LAW— CRITICAL PERSPECTIVES ON CRIME AND THE CRIMINAL PROCESS 53 (2d ed. 1998).

101. See chapters 3 and 4.

102. People v. Monks, 133 Cal. App. 440, 446 (Cal. Dist. Ct. App. 1933).
103. See Solum, *supra* note 68, at pp. 1276–1279.
104. See section 2.3.1.
105. See section 2.3.2.
106. HALLEVY, *supra* note 14, at pp. 241–248.
107. DIGESTA, 48.19.38.5; CODEX JUSTINIANUS, 9.12.6; REINHARD ZIMMERMANN, THE LAW OF OBLIGATIONS — ROMAN FOUNDATIONS OF THE CIVILIAN TRADITION 197 (1996).
108. See, for example, United States v. Greer, 467 F.2d 1064 (7th Cir.1972); People v. Cooper, 194 Ill.2d 419, 252 Ill.Dec. 458, 743 N.E.2d 32 (2000).
109. State v. Lucas, 55 Iowa 321, 7 N.W. 583 (1880); Roy v. United States, 652 A.2d 1098 (D.C.App.1995); People v. Weiss, 256 App.Div. 162, 9 N.Y.S.2d 1 (1939); People v. Little, 41 Cal.App.2d 797, 107 P.2d 634 (1941).
110. State v. Linscott, 520 A.2d 1067 (Me.1987): "a rule allowing for a murder conviction under a theory of accomplice liability based upon an objective standard, despite the absence of evidence that the defendant possessed the culpable subjective mental state that constitutes an element of the crime of murder, does not represent a departure from prior Maine law" (original emphasis).
111. People v. Prettyman, 14 Cal.4th 248, 58 Cal.Rptr.2d 827, 926 P.2d 1013 (1996); Chance v. State, 685 A.2d 351 (Del.1996); Ingram v. United States, 592 A.2d 992 (D.C.App.1991); Richardson v. State, 697 N.E.2d 462 (Ind.1998); Mitchell v. State, 114 Nev. 1417, 971 P.2d 813 (1998); State v. Carrasco, 122 N.M. 554, 928 P.2d 939 (1996); State v. Jackson, 137 Wash.2d 712, 976 P.2d 1229 (1999).
112. United States v. Powell, 929 F.2d 724 (D.C.Cir.1991).
113. State v. Kaiser, 260 Kan. 235, 918 P.2d 629 (1996); United States v. Andrews, 75 F.3d 552 (9th Cir.1996); State v. Goodall, 407 A.2d 268 (Me.1979). Compare: People v. Kessler, 57 Ill.2d 493, 315 N.E.2d 29 (1974).
114. People v. Cabaltero, 31 Cal.App.2d 52, 87 P.2d 364 (1939); People v. Michalow, 229 N.Y. 325, 128 N.E. 228 (1920).
115. Anderson, [1966] 2 Q.B. 110, [1966] 2 All E.R. 644, [1966] 2 W.L.R. 1195, 50 Cr. App. Rep. 216, 130 J.P. 318: "put the principle of law to be invoked in this form: that where two persons embark on a joint enterprise, each is liable for the acts done in pursuance of that joint enterprise, that that includes liability for unusual consequences if they arise from the execution of the agreed joint enterprise."
116. English, [1999] A.C. 1, [1997] 4 All E.R. 545, [1997] 3 W.L.R. 959, [1998] 1 Cr. App. Rep. 261, [1998] Crim. L.R. 48, 162 J.P. 1; Webb, [2006] E.W.C.A. Crim. 2496, [2007] All E.R. (D) 406; O'Flaherty, [2004] E.W.C.A. Crim. 526, [2004] 2 Cr. App. Rep. 315.
117. BGH 24, 213; BGH 26, 176; BGH 26, 244.
118. See section 2.3.2.
119. See section 2.3.1.

3. AI Criminal Liability for Negligence Offenses

1. For the role of AI expert systems in industry and research, see section 1.1.3.

2. See Gabriel Hallevy, *The Criminal Liability of Artificial Intelligence Entities—from Science Fiction to Legal Social Control*, 4 AKRON INTELL. PROP. J. 171 (2010).

3. See section 1.3.2.

4. See section 2.1.

5. See REUVEN YARON, THE LAWS OF ESHNUNNA 264 (2d ed. 1988).

6. COLLATIO MOSAICARUM ET ROMANARUM LEGUM, 1.6.1–4, 1.11.3–4; DIGESTA, 48.8.1.3, 48.19.5.2; ULPIAN, 7 de off. Proconsulis; PAULI SENTENTIAE, 1 manual: "magna neglegentia culpa est; magna culpa dolus est."

7. HENRY DE BRACTON, DE LEGIBUS ET CONSUETUDINIBUS ANGLIAE 278 (1260); G. E. Woodbine ed., S. E. Thorne trans., 1968–1977).

8. Hull, (1664) Kel. 40, 84 E.R. 1072, 1073.

9. Williamson, (1807) 3 Car. & P. 635, 172 E.R. 579.

10. Knight, (1828) 1 L.C.C. 168, 168 E.R. 1000; Grout, (1834) 6 Car. & P. 629, 172 E.R. 1394; Dalloway, (1847) 2 Cox C.C. 273.

11. Finney, (1874) 12 Cox C.C. 625.

12. Bateman, [1925] All E.R. Rep. 45, 94 L.J.K.B. 791, 133 L.T. 730, 89 J.P. 162, 41 T.L.R. 557, 69 Sol. Jo. 622, 28 Cox. C.C. 33, 19 Cr. App. Rep. 8; Leach, [1937] 1 All E.R. 319; Caldwell, [1982] A.C. 341, [1981] 1 All E.R. 961, [1981] 2 W.L.R. 509, 73 Cr. App. Rep. 13, 145 J.P. 211.

13. G., [2003] U.K.H.L. 50, [2003] 4 All E.R. 765, [2004] 1 Cr. App. Rep. 237, 167 J.P. 621, [2004] Crim. L.R. 369, [2004] 1 A.C. 1034.

14. JEROME HALL, GENERAL PRINCIPLES OF CRIMINAL LAW 126 (2d ed. 1960, 2005).

15. Commonwealth v. Thompson, 6 Mass. 134, 6 Tyng 134 (1809); United States v. Freeman, 25 Fed. Cas. 1208 (1827); Rice v. State, 8 Mo. 403 (1844); United States v. Warner, 28 Fed. Cas. 404, 6 W.L.J. 255, 4 McLean 463 (1848); Ann v. State, 30 Tenn. 159, 11 Hum. 159 (1850); State v. Schulz, 55 Ia. 628 (1881).

16. Lee v. State, 41 Tenn. 62, 1 Cold. 62 (1860); Chrystal v. Commonwealth, 72 Ky. 669, 9 Bush. 669 (1873).

17. Commonwealth v. Pierce, 138 Mass. 165 (1884); Abrams v. United States, 250 U.S. 616, 63 L.Ed. 1173, 40 S.Ct. 17 (1919).

18. Commonwealth v. Walensky, 316 Mass. 383, 55 N.E.2d 902 (1944).

19. See, for example, People v. Haney, 30 N.Y.2d 328, 333 N.Y.S.2d 403, 284 N.E.2d 564 (1972); Leet v. State, 595 So.2d 959 (1991); Minor v. State, 326 Md. 436, 605 A.2d 138 (1992); United States v. Hanousek, 176 F.3d 1116 (9th Cir.1999).

20. See, for example, State v. Foster, 91 Wash.2d 466, 589 P.2d 789 (1979); State v. Wilchinski, 242 Conn. 211, 700 A.2d 1 (1997); United States v. Dominguez-Ochoa, 386 F.3d 639 (2004).

21. See, for example, Jerome Hall, *Negligent Behaviour Should Be Excluded from Penal Liability*, 63 COLUM. L. REV. 632 (1963); Robert P. Fine & Gary M. Cohen, *Is Criminal Negligence a Defensible Basis for Criminal Liability?*, 16 BUFF. L. R. 749 (1966).

22. See section 2.1.1.

23. For car accident statistics in the United States, see, for example, http://www.cdc.gov (last visited Feb. 29, 2012).

24. See section 3.2.2.

25. See section 1.3.2.

26. For example, in France (article 121–3 of the French penal code).

27. See section 2.2.2.

28. In Hall v. Brooklands Auto Racing Club, [1932] All E.R. 208, [1933] 1 K.B. 205, 101 L.J.K.B. 679, 147 L.T. 404, 48 T.L.R. 546, the "reasonable person" has been described this way: "The person concerned is sometimes described as 'the man in the street', or 'the man in the Clapham omnibus', or, as I recently read in an American author, 'the man who takes the magazines at home, and in the evening pushes the lawn mower in his shirt sleeves'."

29. State v. Bunkley, 202 Conn. 629, 522 A.2d 795 (1987); State v. Evans, 134 N.H. 378, 594 A.2d 154 (1991).

30. People v. Decina, 2 N.Y.2d 133, 157 N.Y.S.2d 558, 138 N.E.2d 799 (1956); Government of the Virgin Islands v. Smith, 278 F.2d 169 (3rd Cir.1960); People v. Howk, 56 Cal.2d 687, 16 Cal.Rptr. 370, 365 P.2d 426 (1961); State v. Torres, 495 N.W.2d 678 (1993).

31. See section 1.1.2.

32. See section 2.2.2.

33. See section 1.3.

34. See, for example, Rollins v. State, 2009 Ark. 484, 347 S.W.3d 20 (2009); People v. Larkins, 2010 Mich. App. Lexis 1891 (2010); Driver v. State, 2011 Tex. Crim. App. Lexis 4413 (2011).

35. See section 2.3.2.

36. See, for example, Peter Alldridge, *The Doctrine of Innocent Agency*, 2 CRIM. L. F. 45 (1990).

37. See section 2.3.1.

38. See section 2.3.3.

39. See sections 3.3.1 and 3.3.2.

40. See, for example, United States v. Robertson, 33 M.J. 832 (1991); United States v. Buber, 62 M.J. 476 (2006).

41. See section 2.2.2.

4. AI Criminal Liability for Strict Liability Offenses

1. See section 1.1.1.

2. See, for example, W. J. Hennigan, *New Drone Has No Pilot Anywhere, So Who's Accountable?*, LOS ANGELES TIMES, Jan. 26, 2012; see also http://www.latimes.com (last visited Feb. 29, 2012).

3. See section 1.3.2.

4. See section 2.1.

5. Francis Bowes Sayre, *Public Welfare Offenses*, 33 COLUM. L. REV. 55, 56 (1933).

6. See, for example, Nutt, (1728) 1 Barn. K.B. 306, 94 E.R. 208; Dodd, (1736) Sess. Cas. 135, 93 E.R. 136; Almon, (1770) 5 Burr. 2686, 98 E.R. 411; Walter, (1799) 3 Esp. 21, 170 E.R. 524.

7. See, for example, 6 & 7 Vict. c.96.

8. Dixon, (1814) 3 M. & S. 11, 105 E.R. 516; Vantandillo, (1815) 4 M. & S. 73, 105 E.R. 762; Burnett, (1815) 4 M. & S. 272, 105 E.R. 835.

9. Woodrow, (1846) 15 M. & W. 404, 153 E.R. 907.

10. Stephens, [1866] 1 Q.B. 702; Fitzpatrick v. Kelly, [1873] 8 Q.B. 337; Dyke v. Gower, [1892] 1 Q.B. 220; Blaker v. Tillstone, [1894] 1 Q.B. 345; Spiers & Pond v. Bennett, [1896] 2 Q.B. 65; Hobbs v. Winchester Corporation, [1910] 2 K.B. 471; Provincial Motor Cab Company Ltd. v. Dunning, [1909] 2 K.B. 599, 602.

11. W. G. Carson, *Some Sociological Aspects of Strict Liability and the Enforcement of Factory Legislation*, 33 MOD. L. REV. 396 (1970); W. G. Carson, *The Conventionalisation of Early Factory Crime*, 7 INT'L J. OF SOCIOLOGY OF LAW 37 (1979).

12. AUSTIN TURK, CRIMINALITY AND LEGAL ORDER (1969).

13. NICOLA LACEY, CELIA WELLS, & OLIVER QUICK, RECONSTRUCTING CRIMINAL LAW 638–639 (3d ed. 2003, 2006).

14. Barnes v. State, 19 Conn. 398 (1849); Commonwealth v. Boynton, 84 Mass. 160, 2 Allen 160 (1861); Commonwealth v. Goodman, 97 Mass. 117 (1867); Farmer v. People, 77 Ill. 322 (1875); State v. Sasse, 6 S.D. 212, 60 N.W. 853 (1894); State v. Cain, 9 W. Va. 559 (1874); Redmond v. State, 36 Ark. 58 (1880); State v. Clottu, 33 Ind. 409 (1870); State v. Lawrence, 97 N.C. 492, 2 S.E. 367 (1887).

15. Myers v. State, 1 Conn. 502 (1816); Birney v. State, 8 Ohio Rep. 230 (1837); Miller v. State, 3 Ohio St. Rep. 475 (1854); Hunter v. State, 30 Tenn. 160, 1 Head 160 (1858); Stein v. State, 37 Ala. 123 (1861).

16. John R. Spencer & Antje Pedain, *Approaches to Strict and Constructive Liability in Continental Criminal Law*, in APPRAISING STRICT LIABILITY 237 (A. P. Simester ed., 2005).

17. Gammon (Hong Kong) Ltd. v. Attorney-General of Hong Kong, [1985] 1 A.C. 1, [1984] 2 All E.R. 503, [1984] 3 W.L.R. 437, 80 Cr. App. Rep. 194, 26 Build L.R. 159.

18. G., [2003] U.K.H.L. 50, [2003] 4 All E.R. 765, [2004] 1 Cr. App. Rep. 237, 167 J.P. 621, [2004] Crim. L.R. 369; Kumar, [2004] E.W.C.A. Crim. 3207, [2005] 1 Cr. App. Rep. 566, [2005] Crim. L.R. 470; Matudi, [2004] E.W.C.A. Crim. 697.

19. Salibaku v. France, (1998) E.H.R.R. 379.

20. 1950 European Human Rights Covenant, sec. 6(2) provides, "Everyone

charged with a criminal offence shall be presumed innocent until proved guilty according to law."

21. G., [2008] U.K.H.L. 37, [2009] A.C. 92; Barnfather v. Islington London Borough Council, [2003] E.W.H.C. 418 (Admin), [2003] 1 W.L.R. 2318, [2003] E.L.R. 263; G. R. Sullivan, *Strict Liability for Criminal Offences in England and Wales Following Incorporation into English Law of the European Convention on Human Rights, in* APPRAISING STRICT LIABILITY 195 (A. P. Simester ed., 2005).

22. Smith v. California, 361 U.S. 147, 80 S.Ct. 215, 4 L.Ed.2d 205 (1959); Lambert v. California, 355 U.S. 225, 78 S.Ct. 240, 2 L.Ed.2d 228 (1957); Texaco Inc. v. Short, 454 U.S. 516, 102 S.Ct. 781, 70 L.Ed.2d 738 (1982); Carter v. United States, 530 U.S. 255, 120 S.Ct. 2159, 147 L.Ed.2d 203 (2000); Alan C. Michaels, *Imposing Constitutional Limits on Strict Liability: Lessons from the American Experience, in* APPRAISING STRICT LIABILITY 218, 222–223 (A. P. Simester ed., 2005).

23. State v. Stepniewski, 105 Wis.2d 261, 314 N.W.2d 98 (1982); State v. McDowell, 312 N.W.2d 301 (N.D. 1981); State v. Campbell, 536 P.2d 105 (Alaska 1975); Kimoktoak v. State, 584 P.2d 25 (Alaska 1978); Hentzner v. State, 613 P.2d 821 (Alaska 1980); State v. Brown, 389 So.2d 48 (La.1980).

24. B. v. Director of Public Prosecutions, [2000] 2 A.C. 428, [2000] 1 All E.R. 833, [2000] 2 W.L.R. 452, [2000] 2 Cr. App. Rep. 65, [2000] Crim. L.R. 403; Richards, [2004] E.W.C.A. Crim. 192.

25. For the negligence reasonability, see section 3.2.

26. See section 1.3.2.

27. *In re* Welfare of C.R.M., 611 N.W.2d 802 (Minn.2000); State v. Strong, 294 N.W.2d 319 (Minn.1980); Thompson v. State, 44 S.W.3d 171 (Tex. App.2001); State v. Anderson, 141 Wash.2d 357, 5 P.3d 1247 (2000).

28. See section 4.2.1.

29. See sections 2.2.2 and 3.2.2.

30. See section 3.2.1.

31. See section 3.2.

32. See sections 2.2.2 and 3.2.2.

33. See section 1.3.

34. See sections 3.2.1 and 4.2.1.

35. See section 2.3.2.

36. See, for example, Glanville Williams, *Innocent Agency and Causation*, 3 CRIM. L. F. 289 (1992).

37. See section 3.3.3.

38. Not in all legal systems is there a strict liability offense of homicide. In such cases the relevant offense may be the particular traffic offense. See, for example, Dana K. Cole, *Expending Felony-Murder in Ohio: Felony-Murder or Murder-Felony*, 63 OHIO ST. L.J. 15 (2002).

39. See sections 2.2.2 and 3.2.1.

40. See section 3.2.2.

41. See, for example, Peter Lynch, *The Origins of Computer Weather Prediction and Climate Modeling*, 227 JOURNAL OF COMPUTATIONAL PHYSICS 3431 (2008).

5. Applicability of General Defenses to AI Criminal Liability

1. See section 1.1.1.

2. Nick Carbone, South Korea Rolls Out Robotic Prison Guards (Nov. 27, 2011), http://newsfeed.time.com (last visited Jan. 4, 2012); Alex Knapp, South Korean Prison To Feature Robot Guards (Nov. 27, 2011), http://www.forbes.com (last visited Feb. 29, 2012).

3. ANDREW ASHWORTH, PRINCIPLES OF CRIMINAL LAW 157–158, 202 (5th ed. 2006).

4. REUVEN YARON, THE LAWS OF ESHNUNNA 265, 283 (2d ed. 1988).

5. Compare Kent Greenawalt, *Distinguishing Justifications from Excuses*, 49 LAW & CONTEMP. PROBS. 89 (1986); Kent Greenawalt, *The Perplexing Borders of Justification and Excuse*, 84 COLUM. L. REV. 1897 (1984); GEORGE P. FLETCHER, RETHINKING CRIMINAL LAW 759–817 (1978, 2000).

6. Compare Paul H. Robinson, *A Theory of Justification: Societal Harm as a Prerequisite for Criminal Liability*, 23 U.C.L.A. L. REV. 266 (1975); Paul H. Robinson, *Testing Competing Theories of Justification*, 76 N.C. L. REV. 1095 (1998); George P. Fletcher, *The Nature of Justification*, in ACTION AND VALUE IN CRIMINAL LAW 175 (Stephen Shute, John Gardner, & Jeremy Horder eds., 2003).

7. See section 5.1.

8. RUDOLPH SOHM, THE INSTITUTES OF ROMAN LAW 219 (3d ed. 1907).

9. See, for example, Minn. Stat. §9913 (1927); Mont. Rev. Code §10729 (1935); N.Y. Penal Code §816 (1935); Okla. Stat. §152 (1937); Utah Rev. Stat. 103-i-40 (1933).

10. State v. George, 20 Del. 57, 54 A. 745 (1902); Heilman v. Commonwealth, 84 Ky. 457, 1 S.W. 731 (1886); State v. Aaron, 4 N.J.L. 269 (1818).

11. State v. Dillon, 93 Idaho 698, 471 P.2d 553 (1970); State v. Jackson, 346 Mo. 474, 142 S.W.2d 45 (1940).

12. See Godfrey v. State, 31 Ala. 323 (1858); Martin v. State, 90 Ala. 602, 8 So. 858 (1891); State v. J.P.S., 135 Wash.2d 34, 954 P.2d 894 (1998); Beason v. State, 96 Miss. 165, 50 So. 488 (1909); State v. Nickelson, 45 La.Ann. 1172, 14 So. 134 (1893); Commonwealth v. Mead, 92 Mass. 398 (1865); Willet v. Commonwealth, 76 Ky. 230 (1877); Scott v. State, 71 Tex.Crim.R. 41, 158 S.W. 814 (1913); Price v. State, 50 Tex.Crim.R. 71, 94 S.W. 901 (1906).

13. Adams v. State, 8 Md.App. 684, 262 A.2d 69 (1970): "the most modern definition of the test is simply that the surrounding circumstances must demonstrate, beyond a reasonable doubt, that the individual knew what he was doing and that it was wrong."

14. A. W. G. Kean, *The History of the Criminal Liability of Children*, 53 L. Q. REV. 364 (1937).

15. Andrew Walkover, *The Infancy Defense in the New Juvenile Court*, 31 U.C.L.A. L. REV. 503 (1984); Keith Foren, *Casenote: In Re Tyvonne M. Revisited: The Criminal Infancy Defense in Connecticut*, 18 Q. L. REV. 733 (1999).

16. Frederick J. Ludwig, *Rationale of Responsibility for Young Offenders*, 29 NEB. L. REV. 521 (1950); *In re* Tyvonne, 211 Conn. 151, 558 A.2d 661 (1989).

17. See section 5.2.3.

18. See section 5.2.4.

19. In Bratty v. Attorney-General for Northern Ireland, [1963] A.C. 386, 409, [1961] 3 All E.R. 523, [1961] 3 W.L.R. 965, 46 Cr. App. Rep 1, Lord Denning noted, "The requirement that it should be a voluntary act is essential, not only in a murder case, but also in every criminal case. No act is punishable if it is done involuntarily." State v. Mishne, 427 A.2d 450 (Me.1981); State v. Case, 672 A.2d 586 (Me.1996).

20. See, for example, People v. Newton, 8 Cal.App.3d 359, 87 Cal.Rptr. 394 (1970).

21. Kenneth L. Campbell, *Psychological Blow Automatism: A Narrow Defence*, 23 CRIM. L. Q. 342 (1981); Winifred H. Holland, *Automatism and Criminal Responsibility*, 25 CRIM. L. Q. 95 (1982).

22. People v. Higgins, 5 N.Y.2d 607, 186 N.Y.S.2d 623, 159 N.E.2d 179 (1959); State v. Welsh, 8 Wash.App. 719, 508 P.2d 1041 (1973).

23. Reed v. State, 693 N.E.2d 988 (Ind.App.1998).

24. Quick, [1973] Q.B. 910, [1973] 3 All E.R. 347, [1973] 3 W.L.R. 26, 57 Cr. App. Rep. 722, 137 J.P. 763; C, [2007] E.W.C.A. Crim. 1862, [2007] All E.R. (D) 91.

25. Fain v. Commonwealth, 78 Ky. 183 (1879); Bradley v. State, 102 Tex. Crim.R. 41, 277 S.W. 147 (1926); Norval Morris, *Somnambulistic Homicide: Ghosts, Spiders, and North Koreans*, 5 RES JUDICATAE 29 (1951).

26. McClain v. State, 678 N.E.2d 104 (Ind.1997).

27. People v. Newton, 8 Cal.App.3d 359, 87 Cal.Rptr. 394 (1970); Read v. People, 119 Colo. 506, 205 P.2d 233 (1949); Carter v. State, 376 P.2d 351 (Okl. Crim.App.1962).

28. People v. Wilson, 66 Cal.2d 749, 59 Cal.Rptr. 156, 427 P.2d 820 (1967); People v. Lisnow, 88 Cal.App.3d Supp. 21, 151 Cal.Rptr. 621 (1978); Lawrence Taylor & Katharina Dalton, *Premenstrual Syndrome: A New Criminal Defense?*, 19 CAL. W. L. REV. 269 (1983); Michael J. Davidson, *Feminine Hormonal Defenses: Premenstrual Syndrome and Postpartum Psychosis*, 5 ARMY LAWYER (2000).

29. Government of the Virgin Islands v. Smith, 278 F.2d 169 (3rd Cir.1960); People v. Freeman, 61 Cal.App.2d 110, 142 P.2d 435 (1943); State v. Hinkle, 200 W.Va. 280, 489 S.E.2d 257 (1996).

30. State v. Gish, 17 Idaho 341, 393 P.2d 342 (1964); Evans v. State, 322 Md. 24, 585 A.2d 204 (1991); State v. Jenner, 451 N.W.2d 710 (S.D.1990); Lester v. State, 212 Tenn. 338, 370 S.W.2d 405 (1963); Polston v. State, 685 P.2d 1 (Wyo.1984).

31. Richard Delgado, *Ascription of Criminal States of Mind: Toward a Defense Theory for the Coercively Persuaded ("Brainwashed") Defendant*, 63 MINN. L. REV. 1 (1978); Joshua Dressler, *Professor Delgado's "Brainwashing" Defense: Courting a Determinist Legal System*, 63 MINN. L. REV. 335 (1978).

32. FRANCIS ANTONY WHITLOCK, CRIMINAL RESPONSIBILITY AND MENTAL ILLNESS 119–120 (1963).

33. RG 60, 29; RG 73, 177; VRS 23, 212; VRS 46, 440; VRS 61, 339; VRS 64, 189; DAR 1985, 387; BGH 2, 14; BGH 17, 259; BGH 21, 381.

34. KARL MENNINGER, MARTIN MAYMAN, & PAUL PRUYSER, THE VITAL BALANCE 420–489 (1963); George Mora, *Historical and Theoretical Trends in Psychiatry, in* COMPREHENSIVE TEXTBOOK OF PSYCHIATRY 1, 8–19 (Alfred M. Freedman, Harold Kaplan, & Benjamin J. Sadock eds., 2d ed. 1975).

35. MICHAEL MOORE, LAW AND PSYCHIATRY: RETHINKING THE RELATIONSHIP 64–65 (1984); Anthony Platt & Bernard L. Diamond, *The Origins of the "Right and Wrong" Test of Criminal Responsibility and Its Subsequent Development in the United States: An Historical Survey*, 54 CAL. L. REV. 1227 (1966).

36. SANDER L. GILMAN, SEEING THE INSANE (1982); JOHN BIGGS, THE GUILTY MIND 26 (1955).

37. WALTER BROMBERG, FROM SHAMAN TO PSYCHOTHERAPIST: A HISTORY OF THE TREATMENT OF MENTAL ILLNESS 63 (1975); GEORGE ROSEN, MADNESS IN SOCIETY: CHAPTERS IN THE HISTORICAL SOCIOLOGY OF MENTAL ILLNESS 33, 82 (1969); EDWARD NORBECK, RELIGION IN PRIMITIVE SOCIETY 215 (1961).

38. JAMES COWLES PRICHARD, A TREATISE ON INSANITY AND OTHER DISORDERS AFFECTING THE MIND (1835); ARTHUR E. FINK, CAUSES OF CRIME: BIOLOGICAL THEORIES IN THE UNITED STATES, 1800–1915 48–76 (1938); Janet A. Tighe, *Francis Wharton and the Nineteenth Century Insanity Defense: The Origins of a Reform Tradition*, 27 AM. J. LEGAL HIST. 223 (1983).

39. Peter McCandless, *Liberty and Lunacy: The Victorians and Wrongful Confinement, in* MADHOUSES, MAD-DOCTORS, AND MADMEN: THE SOCIAL HISTORY OF PSYCHIATRY IN THE VICTORIAN ERA 339, 354 (Scull ed., 1981); VIEDA SKULTANS, ENGLISH MADNESS: IDEAS ON INSANITY, 1580–1890 69–97 (1979); MICHEL FOUCAULT, MADNESS AND CIVILIZATION 24 (1965).

40. Seymour L. Halleck, *The Historical and Ethical Antecedents of Psychiatric Criminology, in* PSYCHIATRIC ASPECTS OF CRIMINOLOGY 8 (Halleck & Bromberg eds., 1968); FRANZ ALEXANDER & HUGO STAUB, THE CRIMINAL, THE JUDGE, AND THE PUBLIC 24–25 (1931); FRANZ ALEXANDER, OUR AGE OF UNREASON: A STUDY OF THE IRRATIONAL FORCES IN SOCIAL LIFE (rev. ed. 1971).

41. Homer D. Crotty, *The History of Insanity as a Defence to Crime in English Common Law*, 12 CAL. L. REV. 105, 107–108 (1924).

42. M'Naghten, (1843) 10 Cl. & Fin. 200, 8 E.R. 718.

43. Oxford, (1840) 9 Car. & P. 525, 173 E.R. 941.

44. See section 2.2.1.

45. United States v. Freeman, 357 F.2d 606 (2nd Cir.1966); United States

v. Currens, 290 F.2d 751 (3rd Cir.1961); United States v. Chandler, 393 F.2d 920 (4th Cir.1968); Blake v. United States, 407 F.2d 908 (5th Cir.1969); United States v. Smith, 404 F.2d 720 (6th Cir.1968); United States v. Shapiro, 383 F.2d 680 (7th Cir.1967); Pope v. United States, 372 F.2d 710 (8th Cir.1970).

46. Commonwealth v. Herd, 413 Mass. 834, 604 N.E.2d 1294 (1992); State v. Curry, 45 Ohio St.3d 109, 543 N.E.2d 1228 (1989); State v. Barrett, 768 A.2d 929 (R.I.2001); State v. Lockhart, 208 W.Va. 622, 542 S.E.2d 443 (2000). See also 18 U.S.C.A. §17.

47. The American Law Institute, Model Penal Code—Official Draft and Explanatory Notes 61–62 (1962, 1985): "(1) A person is not responsible for criminal conduct if at the time of such conduct as a result of mental disease or defect he lacks substantial capacity either to appreciate the criminality [wrongfulness] of his conduct or to conform his conduct to the requirements of law; (2) As used in this Article, the terms 'mental disease or defect' do not include an abnormality manifested only by repeated criminal or otherwise antisocial conduct."

48. State v. Elsea, 251 S.W.2d 650 (Mo.1952); State v. Johnson, 233 Wis. 668, 290 N.W. 159 (1940); State v. Hadley, 65 Utah 109, 234 P. 940 (1925); HENRY WEIHOFEN, MENTAL DISORDER AS A CRIMINAL DEFENSE 119 (1954); K. W. M. Fulford, *Value, Action, Mental Illness, and the Law*, in ACTION AND VALUE IN CRIMINAL LAW 279 (Stephen Shute, John Gardner, & Jeremy Horder eds., 2003).

49. See sections 2.2.2 and 2.2.3.

50. People v. Sommers, 200 P.3d 1089 (2008); McNeil v. United States, 933 A.2d 354 (2007); Rangel v. State, 2009 Tex.App. 1555 (2009); Commonwealth v. Shumway, 72 Va.Cir. 481 (2007).

51. R. U. Singh, *History of the Defence of Drunkenness in English Criminal Law*, 49 LAW Q. REV. 528 (1933).

52. THEODORI LIBER POENITENTIALIS, III, 13 (668–690).

53. Francis Bowes Sayre, *Mens Rea*, 45 HARV. L. REV. 974, 1014–1015 (1932).

54. WILLIAM OLDNALL RUSSELL, A TREATISE ON CRIMES AND MISDEMEANORS 8 (1843, 1964).

55. Marshall, (1830) 1 Lewin 76, 168 E.R. 965.

56. Pearson, (1835) 2 Lewin 144, 168 E.R. 1108; Thomas, (1837) 7 Car. & P. 817, 173 E.R. 356: "Drunkenness may be taken into consideration in cases where what the law deems sufficient provocation has been given, because the question is, in such cases, whether the fatal act is to be attributed to the passion of anger excited by the previous provocation, and that passion is more easily excitable in a person when in a state of intoxication than when he is sober."

57. Meakin, (1836) 7 Car. & P. 297, 173 E.R. 131; Meade, [1909] 1 K.B. 895; Pigman v. State, 14 Ohio 555 (1846); People v. Harris, 29 Cal. 678 (1866); People v. Townsend, 214 Mich. 267, 183 N.W. 177 (1921).

58. Derrick Augustus Carter, *Bifurcations of Consciousness: The Elimination of the Self-Induced Intoxication Excuse*, 64 MO. L. REV. 383 (1999); Jerome Hall, *Intoxication and Criminal Responsibility*, 57 HARV. L. REV. 1045 (1944); Monrad G. Paulsen, *Intoxication as a Defense to Crime*, 1961 U. ILL. L. F. 1 (1961).

59. State v. Cameron, 104 N.J. 42, 514 A.2d 1302 (1986); State v. Smith, 260 Or. 349, 490 P.2d 1262 (1971); People v. Leonardi, 143 N.Y. 360, 38 N.E. 372 (1894); Tate v. Commonwealth, 258 Ky. 685, 80 S.W.2d 817 (1935); Roberts v. People, 19 Mich. 401 (1870); People v. Kirst, 168 N.Y. 19, 60 N.E. 1057 (1901); State v. Robinson, 20 W.Va. 713, 43 Am.Rep. 799 (1882).

60. Addison M. Bowman, *Narcotic Addiction and Criminal Responsibility under Durham*, 53 GEO. L.J. 1017 (1965); Herbert Fingarette, *Addiction and Criminal Responsibility*, 84 YALE L.J. 413 (1975); Lionel H. Frankel, *Narcotic Addiction, Criminal Responsibility and Civil Commitment*, 1966 UTAH L. REV. 581 (1966); Peter Barton Hutt & Richard A. Merrill, *Criminal Responsibility and the Right to Treatment for Intoxication and Alcoholism*, 57 GEO. L.J. 835 (1969).

61. Powell v. Texas, 392 U.S. 514, 88 S.Ct. 2145, 20 L.Ed.2d 1254 (1968); United States v. Moore, 486 F.2d 1139 (D.C.Cir.1973); State v. Herro, 120 Ariz. 604, 587 P.2d 1181 (1978); State v. Smith, 219 N.W.2d 655 (Iowa 1974); People v. Davis, 33 N.Y.2d 221, 351 N.Y.S.2d 663, 306 N.E.2d 787 (1973).

62. See sections 2.2.2 and 2.2.3.

63. See sections 2.2.1 and 2.2.2.

64. William G. Lycan, *Introduction, in* MIND AND COGNITION 3–13 (William G. Lycan ed., 1990).

65. Edwin R. Keedy, *Ignorance and Mistake in the Criminal Law*, 22 HARV. L. REV. 75, 78 (1909); Levett, (1638) Cro. Car. 538.

66. State v. Silveira, 198 Conn. 454, 503 A.2d 599 (1986); State v. Molin, 288 N.W.2d 232 (Minn.1979); State v. Sexton, 160 N.J. 93, 733 A.2d 1125 (1999).

67. State v. Sawyer, 95 Conn. 34, 110 A. 461 (1920); State v. Cude, 14 Utah 2d 287, 383 P.2d 399 (1963); Ratzlaf v. United States, 510 U.S. 135, 114 S.Ct. 655, 126 L.Ed.2d 615 (1994); Cheek v. United States, 498 U.S. 192, 111 S.Ct. 604, 112 L.Ed.2d 617 (1991); Richard H. S. Tur, *Subjectivism and Objectivism: Towards Synthesis, in* ACTION AND VALUE IN CRIMINAL LAW 213 (Stephen Shute, John Gardner, & Jeremy Horder eds., 2003).

68. United States v. Lampkins, 4 U.S.C.M.A. 31, 15 C.M.R. 31 (1954).

69. People v. Vogel, 46 Cal.2d 798, 299 P.2d 850 (1956); Long v. State, 44 Del. 262, 65 A.2d 489 (1949).

70. Fernand N. Dutile & Harold F. Moore, *Mistake and Impossibility: Arranging Marriage Between Two Difficult Partners*, 74 NW. U. L. REV. 166 (1980).

71. Douglas Husak & Andrew von Hirsch, *Culpability and Mistake of Law, in* ACTION AND VALUE IN CRIMINAL LAW 157, 161–167 (Stephen Shute, John Gardner, & Jeremy Horder eds., 2003).

72. DIGESTA, 22.6.9: "juris quidam ignorantiam cuique nocere, facti vero ignorantiam non nocere."

73. See, for example, Brett v. Rigden, (1568) 1 Plowd. 340, 75 E.R. 516; Mildmay, (1584) 1 Co. Rep. 175a, 76 E.R. 379; Manser, (1584) 2 Co. Rep. 3, 76 E.R. 392; Vaux, (1613) 1 Blustrode 197, 80 E.R. 885; Bailey, (1818) Russ. & Ry. 341, 168 E.R. 835; Esop, (1836) 7 Car. & P. 456, 173 E.R. 203; Crawshaw, (1860) Bell. 303, 169 E.R. 1271; Schuster v. State, 48 Ala. 199 (1872).

74. Forbes, (1835) 7 Car. & P. 224, 173 E.R. 99; Parish, (1837) 8 Car. & P. 94, 173 E.R. 413; Allday, (1837) 8 Car. & P. 136, 173 E.R. 431; Dotson v. State, 6 Cold. 545 (1869); Cutter v. State, 36 N.J.L. 125 (1873); Squire v. State, 46 Ind. 459 (1874).

75. State v. Goodenow, 65 Me. 30 (1876); State v. Whitoomb, 52 Iowa 85, 2 N.W. 970 (1879).

76. Lutwin v. State, 97 N.J.L. 67, 117 A. 164 (1922); State v. Whitman, 116 Fla. 196, 156 So. 705 (1934); United States v. Mancuso, 139 F.2d 90 (3rd Cir.1943); State v. Chicago, M. & St.P.R. Co., 130 Minn. 144, 153 N.W. 320 (1915); Coal & C.R. v. Conley, 67 W.Va. 129, 67 S.E. 613 (1910); State v. Striggles, 202 Iowa 1318, 210 N.W. 137 (1926); United States v. Albertini, 830 F.2d 985 (9th Cir.1987).

77. State v. Sheedy, 125 N.H. 108, 480 A.2d 887 (1984); People v. Ferguson, 134 Cal.App. 41, 24 P.2d 965 (1933); Andrew Ashworth, *Testing Fidelity to Legal Values: Official Involvement and Criminal Justice*, 63 MOD. L. REV. 663 (2000); Glanville Williams, *The Draft Code and Reliance upon Official Statements*, 9 LEGAL STUD. 177 (1989).

78. Rollin M. Perkins, *Ignorance and Mistake in Criminal Law*, 88 U. PA. L. REV. 35 (1940).

79. See section 4.2.2.

80. See sections 4.3.2 and 4.3.3.

81. See section 1.1.1.

82. See section 5.1.

83. Chas E. George, *Limitation of Police Powers*, 12 LAW. & BANKER & S. BENCH & B. REV. 740 (1919); Kam C. Wong, *Police Powers and Control in the People's Republic of China: The History of Shoushen*, 10 COLUM. J. ASIAN L. 367 (1996); John S. Baker, Jr., *State Police Powers and the Federalization of Local Crime*, 72 TEMP. L. REV. 673 (1999).

84. Dolores A. Donovan & Stephanie M. Wildman, *Is the Reasonable Man Obsolete? A Critical Perspective on Self-Defense and Provocation*, 14 LOY. L. A. L. REV. 435, 441 (1981); Joshua Dressler, *Rethinking Heat of Passion: A Defense in Search of a Rationale*, 73 J. CRIM. L. & CRIMINOLOGY 421, 444–450 (1982); Kent Greenawalt, *The Perplexing Borders of Justification and Excuse*, 84 COLUM. L. REV. 1897, 1898, 1915–1919 (1984).

85. State v. Brosnan, 221 Conn. 788, 608 A.2d 49 (1992); State v. Gallagher, 191 Conn. 433, 465 A.2d 323 (1983); State v. Nelson, 329 N.W.2d 643 (Iowa 1983); State v. Farley, 225 Kan. 127, 587 P.2d 337 (1978).

86. Commonwealth v. Monico, 373 Mass. 298, 366 N.E.2d 1241 (1977); Commonwealth v. Johnson, 412 Mass. 368, 589 N.E.2d 311 (1992); Duckett v. State, 966 P.2d 941 (Wyo.1998); People v. Young, 11 N.Y.2d 274, 229 N.Y.S.2d 1, 183 N.E.2d 319 (1962); Batson v. State, 113 Nev. 669, 941 P.2d 478 (1997); State v. Wenger, 58 Ohio St.2d 336, 390 N.E.2d 801 (1979); Moore v. State, 25 Okl.Crim. 118, 218 P. 1102 (1923).

87. Williams v. State, 70 Ga.App. 10, 27 S.E.2d 109 (1943); State v. Totman, 80 Mo.App. 125 (1899).

88. Lawson, [1986] V.R. 515; Daniel v. State, 187 Ga. 411, 1 S.E.2d 6 (1939).

89. John Barker Waite, *The Law of Arrest*, 24 TEX. L. REV. 279 (1946).

90. People v. Williams, 56 Ill.App.2d 159, 205 N.E.2d 749 (1965); People v. Minifie, 13 Cal.4th 1055, 56 Cal.Rptr.2d 133, 920 P.2d 1337 (1996); State v. Coffin, 128 N.M. 192, 991 P.2d 477 (1999).

91. State Philbrick, 402 A.2d 59 (Me.1979); State v. Havican, 213 Conn. 593, 569 A.2d 1089 (1990); State v. Harris, 222 N.W.2d 462 (Iowa 1974); Judith Fabricant, *Homicide in Response to a Threat of Rape: A Theoretical Examination of the Rule of Justification*, 11 GOLDEN GATE U. L. REV. 945 (1981).

92. Celia Wells, *Battered Woman Syndrome and Defences to Homicide: Where Now?*, 14 LEGAL STUD. 266 (1994); Aileen McColgan, *In Defence of Battered Women who Kill*, 13 OXFORD J. LEGAL STUD. 508 (1993); Joshua Dressler, *Battered Women Who Kill Their Sleeping Tormenters: Reflections on Maintaining Respect for Human Life while Killing Moral Monsters*, in CRIMINAL LAW THEORY — DOCTRINES OF THE GENERAL PART 259 (Stephen Shute & A. P. Simester eds., 2005).

93. State v. Moore, 158 N.J. 292, 729 A.2d 1021 (1999); State v. Robinson, 132 Ohio App.3d 830, 726 N.E.2d 581 (1999).

94. Lawrence B. Solum, *Legal Personhood for Artificial Intelligences*, 70 N.C. L. REV. 1231, 1255–1258 (1992).

95. See section 1.2.1; ISAAC ASIMOV, I, ROBOT 40 (1950).

96. See, for example, United States v. Allegheny Bottling Company, 695 F.Supp. 856 (1988); John C. Coffee, Jr., *"No Soul to Damn: No Body to Kick": An Unscandalised Inquiry into the Problem of Corporate Punishment*, 79 MICH. L. REV. 386 (1981).

97. JUDITH JARVIS THOMSON, RIGHTS, RESTITUTION AND RISK: ESSAYS IN MORAL THEORY 33–48 (1986); Sanford Kadish, *Respect for Life and Regard for Rights in the Criminal Law*, 64 CAL. L. REV. 871 (1976); Patrick Montague, *Self-Defense and Choosing Between Lives*, 40 PHIL. STUD. 207 (1981); Cheyney C. Ryan, *Self-Defense, Pacifism, and the Possibility of Killing*, 93 ETHICS 508 (1983).

98. See section 1.2.1.

99. See section 1.1.1.

100. See section 3.2.2.

101. See section 5.3.1.

102. See, for example, United States v. Holmes, 26 F. Cas. 360, 1 Wall. Jr. 1 (1842); Dudley and Stephens, [1884] 14 Q.B. D. 273.

103. W. H. Hitchler, *Necessity as a Defence in Criminal Cases*, 33 DICK. L. REV. 138 (1929).

104. Edward B. Arnolds & Norman F. Garland, *The Defense of Necessity in Criminal Law: The Right to Choose the Lesser Evil*, 65 J. CRIM. L. & CRIMINOLOGY 289 (1974); Lawrence P. Tiffany & Carl A. Anderson, *Legislating the Necessity Defense in Criminal Law*, 52 DENV. L.J. 839 (1975); Rollin M. Perkins, *Impelled Perpetration Restated*, 33 HASTINGS L.J. 403 (1981).

105. Long v. Commonwealth, 23 Va.App. 537, 478 S.E.2d 324 (1996); State v. Crocker, 506 A.2d 209 (Me.1986); Humphrey v. Commonwealth, 37 Va.App. 36, 553 S.E.2d 546 (2001); United States v. Oakland Cannabis Buyers' Cooperative, 532 U.S. 483, 121 S.Ct. 1711, 149 L.Ed.2d 722 (2001); United States v. Kabat, 797 F.2d 580 (8th Cir.1986); McMillan v. City of Jackson, 701 So.2d 1105 (Miss.1997).

106. BENJAMIN THORPE, ANCIENT LAWS AND INSTITUTES OF ENGLAND 47–49 (1840, 2004); Reniger v. Fogossa, (1551) 1 Plowd. 1, 75 E.R. 1, 18; Mouse, (1608) 12 Co. Rep. 63, 77 E.R. 1341; MICHAEL DALTON, THE COUNTREY JUSTICE ch. 150 (1618, 2003).

107. United States v. Randall, 104 Wash.D.C.Rep. 2249 (D.C.Super.1976); State v. Hastings, 118 Idaho 854, 801 P.2d 563 (1990); People v. Whipple, 100 Cal.App. 261, 279 P. 1008 (1929); United States v. Paolello, 951 F.2d 537 (3rd Cir.1991).

108. Commonwealth v. Weaver, 400 Mass. 612, 511 N.E.2d 545 (1987); Nelson v. State, 597 P.2d 977 (Alaska 1979); City of Chicago v. Mayer, 56 Ill.2d 366, 308 N.E.2d 601 (1974); State v. Kee, 398 A.2d 384 (Me.1979); State v. Caswell, 771 A.2d 375 (Me.2001); State v. Jacobs, 371 So.2d 801 (La.1979); Anthony M. Dillof, *Unraveling Unknowing Justification*, 77 NOTRE DAME L. REV. 1547 (2002).

109. United States v. Contento-Pachon, 723 F.2d 691 (9th Cir.1984); United States v. Bailey, 444 U.S. 394, 100 S.Ct. 624, 62 L.Ed.2d 575 (1980); Hunt v. State, 753 So.2d 609 (Fla.App.2000); State v. Anthuber, 201 Wis.2d 512, 549 N.W.2d 477 (App.1996).

110. State v. Fee, 126 N.H. 78, 489 A.2d 606 (1985); United States v. Sued-Jimenez, 275 F.3d 1 (1st Cir.2001); United States v. Dorrell, 758 F.2d 427 (9th Cir.1985); State v. Marley, 54 Haw. 450, 509 P.2d 1095 (1973); State v. Dansinger, 521 A.2d 685 (Me.1987); State v. Champa, 494 A.2d 102 (R.I.1985); Wilson v. State, 777 S.W.2d 823 (Tex.App.1989); State v. Cram, 157 Vt. 466, 600 A.2d 733 (1991).

111. United States v. Maxwell, 254 F.3d 21 (1st Cir.2001); Andrews v. People, 800 P.2d 607 (Colo.1990); State v. Howley, 128 Idaho 874, 920 P.2d 391 (1996); State v. Dansinger, 521 A.2d 685 (Me.1987); Commonwealth v. Leno, 415 Mass. 835, 616 N.E.2d 453 (1993); Commonwealth v. Lindsey,

396 Mass. 840, 489 N.E.2d 666 (1986); People v. Craig, 78 N.Y.2d 616, 578 N.Y.S.2d 471, 585 N.E.2d 783 (1991); State v. Warshow, 138 Vt. 22, 410 A.2d 1000 (1979).

112. See section 5.3.1.

113. *Id.*

114. John Lawrence Hill, *A Utilitarian Theory of Duress*, 84 IOWA L. REV. 275 (1999); Rollin M. Perkins, *Impelled Perpetration Restated*, 33 HASTINGS L.J. 403 (1981); United States v. Johnson, 956 F.2d 894 (9th Cir.1992); Sanders v. State, 466 N.E.2d 424 (Ind.1984); State v. Daoud, 141 N.H. 142, 679 A.2d 577 (1996); Alford v. State, 866 S.W.2d 619 (Tex.Crim.App.1993).

115. McGrowther, (1746) 18 How. St. Tr. 394.

116. United States v. LaFleur, 971 F.2d 200 (9th Cir.1991); Hunt v. State, 753 So.2d 609 (Fla.App.2000); Taylor v. State, 158 Miss. 505, 130 So. 502 (1930); State v. Finnell, 101 N.M. 732, 688 P.2d 769 (1984); State v. Nargashian, 26 R.I. 299, 58 A. 953 (1904); State v. Rocheville, 310 S.C. 20, 425 S.E.2d 32 (1993); Arp v. State, 97 Ala. 5, 12 So. 301 (1893).

117. State v. Nargashian, 26 R.I. 299, 58 A. 953 (1904).

118. People v. Merhige, 212 Mich. 601, 180 N.W. 418 (1920); People v. Pantano, 239 N.Y. 416, 146 N.E. 646 (1925); Tully v. State, 730 P.2d 1206 (Okl.Crim.App.1986); Pugliese v. Commonwealth, 16 Va.App. 82, 428 S.E.2d 16 (1993).

119. United States v. Bakhtiari, 913 F.2d 1053 (2nd Cir.1990); R.I. Recreation Center v. Aetna Cas. & Surety Co., 177 F.2d 603 (1st Cir.1949); Sam v. Commonwealth, 13 Va.App. 312, 411 S.E.2d 832 (1991).

120. Commonwealth v. Perl, 50 Mass.App.Ct. 445, 737 N.E.2d 937 (2000); United States v. Contento-Pachon, 723 F.2d 691 (9th Cir.1984); State v. Ellis, 232 Or. 70, 374 P.2d 461 (1962); State v. Torphy, 78 Mo.App. 206 (1899).

121. People v. Richards, 269 Cal.App.2d 768, 75 Cal.Rptr. 597 (1969); United States v. Bailey, 444 U.S. 394, 100 S.Ct. 624, 62 L.Ed.2d 575 (1980); United States v. Gomez, 81 F.3d 846 (9th Cir.1996); United States v. Arthurs, 73 F.3d 444 (1st Cir.1996); United States v. Lee, 694 F.2d 649 (11th Cir.1983); United States v. Campbell, 675 F.2d 815 (6th Cir.1982); State v. Daoud, 141 N.H. 142, 679 A.2d 577 (1996).

122. United States v. Bailey, 444 U.S. 394, 100 S.Ct. 624, 62 L.Ed.2d 575 (1980); People v. Handy, 198 Colo. 556, 603 P.2d 941 (1979); State v. Reese, 272 N.W.2d 863 (Iowa 1978); State v. Reed, 205 Neb. 45, 286 N.W.2d 111 (1979).

123. Fitzpatrick, [1977] N.I. 20; Hasan, [2005] U.K.H.L. 22, [2005] 4 All E.R. 685, [2005] 2 Cr. App. Rep. 314, [2006] Crim. L.R. 142, [2005] All E.R. (D) 299.

124. See sections 5.3.1 and 5.3.2.

125. Michael A. Musmanno, *Are Subordinate Officials Penally Responsible for Obeying Superior Orders which Direct Commission of Crime?*, 67 DICK. L. REV. 221 (1963).

126. Axtell, (1660) 84 E.R. 1060; Calley v. Callaway, 519 F.2d 184 (5th Cir.1975); United States v. Calley, 48 C.M.R. 19, 22 U.S.C.M.A. 534 (1973).

127. A. P. ROGERS, LAW ON THE BATTLEFIELD 143–147 (1996).

128. Jurco v. State, 825 P.2d 909 (Alaska App.1992); State v. Stoehr, 134 Wis.2d 66, 396 N.W.2d 177 (1986).

129. For the machine learning, see section 1.1.2.

130. See, for example, sections 5.3.1, 5.3.2, and 5.3.3.

131. Vashon R. Rogers, Jr., *De Minimis Non Curat Lex*, 21 ALBANY L.J. 186 (1880); Max L. Veech & Charles R. Moon, *De Minimis non Curat Lex*, 45 MICH. L. REV. 537 (1947).

132. The American Law Institute, Model Penal Code—Official Draft and Explanatory Notes 40 (1962, 1985).

133. Stanislaw Pomorski, *On Multiculturalism, Concepts of Crime, and the "De Minimis" Defense*, 1997 B.Y.U. L. REV. 51 (1997).

6. Sentencing AI

1. BRONISLAW MALINOWSKI, CRIME AND CUSTOM IN SAVAGE SOCIETY (1959, 1982).

2. EXODUS 21:24.

3. GERTRUDE EZORSKY, PHILOSOPHICAL PERSPECTIVES ON PUNISHMENT 102–134 (1972).

4. Sheldon Glueck, *Principles of a Rational Penal Code*, 41 HARV. L. REV. 453 (1928).

5. Jackson Toby, *Is Punishment Necessary?*, 55 J. CRIM. L. CRIMINOLOGY & POLICE SCI. 332 (1964); C. G. Schoenfeld, *In Defence of Retribution in the Law*, 35 PSYCHOANALYTIC Q. 108 (1966).

6. FRANCIS ALLEN, THE DECLINE OF THE REHABILITATIVE IDEAL 66 (1981).

7. BARBARA HUDSON, UNDERSTANDING JUSTICE: AN INTRODUCTION TO IDEAS, PERSPECTIVES AND CONTROVERSIES IN MODERN PENAL THEORY 39 (1996, 2003); JESSICA MITFORD, KIND AND USUAL PUNISHMENT: THE PRISON BUSINESS (1974).

8. Russell L. Christopher, *Deterring Retributivism: The Injustice of "Just" Punishment*, 96 NW. U. L. REV. 843 (2002); Douglas Husak, *Holistic Retribution*, 88 CAL. L. REV. 991 (2000); Douglas Husak, *Retribution in Criminal Theory*, 37 SAN DIEGO L. REV. 959 (2000); Dan Markel, *Are Shaming Punishments Beautifully Retributive? Retributivism and the Implications for the Alternative Sanctions Debate*, 54 VAND. L. REV. 2157 (2001).

9. ANDREW VON HIRSCH, DOING JUSTICE: THE CHOICE OF PUNISHMENT 74–75 (1976); Andrew von Hirsch, Proportionate Sentences: A Desert Perspective, in PRINCIPLED SENTENCING: READINGS ON THEORY AND POLICY 115 (Andrew von Hirsch, Andrew Ashworth, & Julian Roberts eds., 3d ed. 2009).

10. Paul H. Robinson & John M. Darley, *The Utility of Desert*, 91 NW. U. L. REV. 453 (1997); Samuel Scheffler, *Justice and Desert in Liberal Theory*, 88 CAL. L. REV. 965 (2000); Edward M. Wise, *The Concept of Desert*, 33 WAYNE L. REV. 1343 (1987).

11. CESARE BECCARIA, TRAITÉ DES DÉLITS ET DES PEINES (1764).

12. LEON RADZINOWICZ, A HISTORY OF ENGLISH CRIMINAL LAW AND ITS ADMINISTRATION FROM 1750 VOL. 1: THE MOVEMENT FOR REFORM (1948); JEREMY BENTHAM, AN INTRODUCTION TO THE PRINCIPLES OF MORALS AND LEGISLATION ch. 13 (1789, 1996); Jeremy Bentham, *Punishment and Deterrence, in* PRINCIPLED SENTENCING: READINGS ON THEORY AND POLICY 53–56 (Andrew von Hirsch, Andrew Ashworth, & Julian Roberts eds., 3d ed. 2009); PAUL JOHANN ANSELM FEUERBACH, LEHRBUCH DES GEMEINEN IN DEUTSCHLAND GÜLTIGEN PEINLICHEN RECHTS 117 (1812, 2007).

13. Johannes Andenaes, *The General Preventive Effects of Punishment*, 114 U. PA. L. REV. 949 (1966); Johannes Andenaes, *The Morality of Deterrence*, 37 U. CHI. L. REV. 649 (1970).

14. JAMES Q. WILSON, THINKING ABOUT CRIME 123–142 (2d ed. 1985); Laurence H. Ross, *Deterrence Regained: The Cheshire Constabulary's "Breathalyser Blitz,"* 6 J. LEGAL STUD. 241 (1977).

15. Gabriel Hallevy, *The Recidivist Wants to Be Punished—Punishment as an Incentive to Re-offend*, 5 INT'L J. PUNISHMENT & SENTENCING 124 (2009).

16. LEON RADZINOWICZ & ROGER HOOD, A HISTORY OF ENGLISH CRIMINAL LAW AND ITS ADMINISTRATION FROM 1750 VOL. 5: THE EMERGENCE OF PENAL POLICY 8–57 (1986).

17. Andrew Ashworth, *Rehabilitation, in* PRINCIPLED SENTENCING: READINGS ON THEORY AND POLICY 1, 1–10 (Andrew von Hirsch, Andrew Ashworth, & Julian Roberts eds., 3d ed. 2009).

18. Robert Martinson, *What Works? Questions and Answers about Prison Reform*, 35 PUBLIC INTEREST 22 (1974).

19. DOUGLAS S. LIPTON, ROBERT MARTINSON, & JUDITH WILKS, THE EFFECTIVENESS OF CORRECTIONAL TREATMENT: A SURVEY OF TREATMENT EVALUATION STUDIES (1975).

20. LEE SECHREST, SUSAN O. WHITE, & ELIZABETH D. BROWN, THE REHABILITATION OF CRIMINAL OFFENDERS: PROBLEMS AND PROSPECTS 27–34 (1979); David F. Greenberg, *The Corrective Effects of Corrections: A Survey of Evaluation, in* CORRECTIONS AND PUNISHMENT 111 (David F. Greenberg ed., 1977).

21. See, for example, MICHEL FOUCAULT, DISCIPLINE AND PUNISH: THE BIRTH OF THE PRISON (1977); ANTHONY M. PLATT, THE CHILD SAVERS: THE INVENTION OF DELINQUENCY (2d ed. 1969, 1977); FRANCIS ALLEN, THE DECLINE OF THE REHABILITATIVE IDEAL 66 (1981); A CLOCKWORK ORANGE (Warner Bros. 1971).

22. Gabriel Hallevy, *Therapeutic Victim-Offender Mediation within the Criminal Justice Process—Sharpening the Evaluation of Personal Potential for Rehabilitation while Righting Wrongs under the Alternative-Dispute-Resolution (ADR) Philosophy*, 16 HARV. NEGOT. L. REV. 65 (2011).

23. DAVID P. FARRINGTON & BRANDON C. WELSH, PREVENTING CRIME: WHAT WORKS FOR CHILDREN, OFFENDERS, VICTIMS AND PLACES (2006); LAWRENCE W. SHERMAN, DAVID P. FARRINGTON, DORIS LEYTON MACKENZIE, & BRANDON C.

WELSH, EVIDENCE-BASED CRIME PREVENTION (2006); ROSEMARY SHEEHAN, GILL MCLVOR, & CHRIS TROTTER, WHAT WORKS WITH WOMEN OFFENDERS (2007); Laaman v. Helgemoe, 437 F.Supp. 269 (1977).

24. Richard P. Seiter & Karen R. Kadela, *Prisoner Reentry: What Works,What Does Not, and What Is Promising*, 49 CRIME & DELINQUENCY 360 (2003); Clive R. Hollin, *Treatment Programs for Offenders*, 22 INT'L J. OF LAW & PSYCHIATRY 361 (1999).

25. See section 3.2.2.

26. DAVID GARLAND, THE CULTURE OF CONTROL: CRIME AND SOCIAL ORDER IN CONTEMPORARY SOCIETY 102 (2002).

27. Ledger Wood, *Responsibility and Punishment*, 28 AM. INST. CRIM. L. & CRIMINOLOGY 630, 639 (1938).

28. Martin P. Kafka, *Sex Offending and Sexual Appetite: The Clinical and Theoretical Relevance of Hypersexual Desire*, 47 INT'L J. OF OFFENDER THERAPY & COMP. CRIMINOLOGY 439 (2003); Matthew Jones, *Overcoming the Myth of Free Will in Criminal Law: The True Impact of the Genetic Revolution*, 52 DUKE L.J. 1031 (2003); Sanford H. Kadish, *Excusing Crime*, 75 CAL. L. REV. 257 (1987).

29. See section 1.3.3.

30. Stuart Field & Nico Jorg, *Corporate Liability and Manslaughter: Should We Be Going Dutch?*, [1991] CRIM. L.R. 156 (1991).

31. Gerard E. Lynch, *The Role of Criminal Law in Policing Corporate Misconduct*, 60 LAW & CONTEMP. PROBS. 23 (1997); Richard Gruner, *To Let the Punishment Fit the Organization: Sanctioning Corporate Offenders Through Corporate Probation*, 16 AM. J. CRIM. L. 1 (1988); Steven Walt & William S. Laufer, *Why Personhood Doesn't Matter: Corporate Criminal Liability and Sanctions*, 18 AM. J. CRIM. L. 263 (1991).

32. United States v. Allegheny Bottling Company, 695 F.Supp. 856 (1988).

33. *Id.* at p. 858.

34. *Id.*

35. John C. Coffee, Jr., *"No Soul to Damn: No Body to Kick": An Unscandalised Inquiry Into the Problem of Corporate Punishment*, 79 MICH. L. REV. 386 (1981); STEVEN BOX, POWER, CRIME AND MYSTIFICATION 16–79 (1983); Brent Fisse & John Braithwaite, *The Allocation of Responsibility for Corporate Crime: Individualism, Collectivism and Accountability*, 11 SYDNEY L. REV. 468 (1988).

36. Allegheny Bottling Company case, *supra* note 32, at p. 861.

37. *Id.* at p. 861.

38. RUSS VERSTEEG, EARLY MESOPOTAMIAN LAW 126 (2000); G. R. DRIVER & JOHN C. MILES, THE BABYLONIAN LAWS, VOL. I: LEGAL COMMENTARY 206, 495–496 (1952): "The capital penalty is most often expressed by saying that the offender 'shall be killed' . . . ; this occurs seventeen times in the first thirty-four

sections. A second form of expression, which occurs five times, is that 'they shall kill' . . . the offender."

39. Frank E. Hartung, *Trends in the Use of Capital Punishment*, 284(1) ANNALS AM. ACAD. POL. & SOC. SCI. 8 (1952).

40. Gregg v. Georgia, 428 U.S. 153, S.Ct. 2909, 49 L.Ed.2d 859 (1979).

41. Provenzano v. Moore, 744 So.2d 413 (Fla. 1999); Dutton v. State, 123 Md. 373, 91 A. 417 (1914); Campbell v. Wood, 18 F.3d 662 (9th Cir. 1994); Wilkerson v. Utah, 99 U.S. (9 Otto) 130, 25 L.Ed. 345 (1878); People v. Daugherty, 40 Cal.2d 876, 256 P.2d 911 (1953); Gray v. Lucas, 710 F.2d 1048 (5th Cir. 1983); Hunt v. Nuth, 57 F.3d 1327 (4th Cir. 1995).

42. ROBERT M. BOHM, DEATHQUEST: AN INTRODUCTION TO THE THEORY AND PRACTICE OF CAPITAL PUNISHMENT IN THE UNITED STATES 74 (1999).

43. Peter Fitzpatrick, *"Always More to Do": Capital Punishment and the (De)Composition of Law*, in THE KILLING STATE — CAPITAL PUNISHMENT IN LAW, POLITICS, AND CULTURE 117 (Austin Sarat ed., 1999); Franklin E. Zimring, *The Executioner's Dissonant Song: On Capital Punishment and American Legal Values*, in THE KILLING STATE — CAPITAL PUNISHMENT IN LAW, POLITICS, AND CULTURE 137 (Austin Sarat ed., 1999).

44. Anne Norton, *After the Terror: Mortality, Equality, Fraternity*, in THE KILLING STATE — CAPITAL PUNISHMENT IN LAW, POLITICS, AND CULTURE 27 (Austin Sarat ed., 1999); Hugo Adam Bedau, *Abolishing the Death Penalty Even for the Worst Murderers*, in THE KILLING STATE — CAPITAL PUNISHMENT IN LAW, POLITICS, AND CULTURE 40 (Austin Sarat ed., 1999).

45. Sean McConville, *The Victorian Prison: England 1865–1965*, in THE OXFORD HISTORY OF THE PRISON 131 (Norval Morris & David J. Rothman eds., 1995); THORSTEN J. SELLIN, SLAVERY AND THE PENAL SYSTEM (1976); HORSFALL J. TURNER, THE ANNALS OF THE WAKEFIELD HOUSE OF CORRECTIONS FOR THREE HUNDRED YEARS 154–172 (1904).

46. JOHN HOWARD, THE STATE OF PRISONS IN ENGLAND AND WALES (1777, 1996).

47. David J. Rothman, *For the Good of All: The Progressive Tradition in Prison Reform*, in HISTORY AND CRIME: IMPLICATIONS FOR CRIMINAL JUSTICE POLICY 271 (James A. Inciardi & Charles E. Faupel eds., 1980).

48. Roy D. King, *The Rise and Rise of Supermax: An American Solution in Search of a Problem?*, 1 PUNISHMENT & SOCIETY 163 (1999); CHASE RIVELAND, SUPERMAX PRISONS: OVERVIEW AND GENERAL CONSIDERATIONS (1999); JAMIE FELLNER & JOANNE MARINER, COLD STORAGE: SUPER-MAXIMUM SECURITY CONFINEMENT IN INDIANA (1997).

49. DORRIS LAYTON MACKANZIE & EUGENE E. HEBERT, CORRECTIONAL BOOT CAMPS: A TOUGH INTERMEDIATE SANCTION (1996); Sue Frank, *Oklahoma Camp Stresses Structure and Discipline*, 53 CORRECTIONS TODAY 102 (1991); ROBERTA C. CRONIN, BOOT CAMPS FOR ADULT AND JUVENILE OFFENDERS: OVERVIEW AND UPDATE (1994).

50. See section 6.1.1.

51. MARC ANCEL, SUSPENDED SENTENCE 14–17 (1971); Marc Ancel, *The System of Conditional Sentence or Sursis*, 80 L. Q. REV. 334, 336 (1964).

52. United Nations, PROBATION AND RELATED MEASURES, UN Department of Social Affairs 29–30 (1951).

53. DAVID J. ROTHMAN, CONSCIENCE AND CONVENIENCE: THE ASYLUM AND ITS ALTERNATIVES IN PROGRESSIVE AMERICA (1980); FRANK SCHMALLEGER, CRIMINAL JUSTICE TODAY: AN INTRODUCTORY TEXT FOR THE 21ST CENTURY 454 (2003).

54. Paul W. Keve, *The Professional Character of the Presentence Report*, 26 FED. PROBATION 51 (1962).

55. HARRY E. ALLEN, ERIC W. CARLSON, & EVALYN C. PARKS, CRITICAL ISSUES IN ADULT PROBATION (1979); Crystal A. Garcia, *Using Palmer's Global Approach to Evaluate Intensive Supervision Programs: Implications for Practice*, 4 CORRECT. MANAG. Q. 60 (2000); ANDREW WRIGHT, GWYNETH BOSWELL, & MARTIN DAVIES, CONTEMPORARY PROBATION PRACTICE (1993); MICHAEL CAVADINO & JAMES DIGNAN, THE PENAL SYSTEM: AN INTRODUCTION 137–140 (2002).

56. John Harding, *The Development of the Community Service*, in ALTERNATIVE STRATEGIES FOR COPING WITH CRIME 164 (Norman Tutt ed., 1978); Home Office, REVIEW OF CRIMINAL JUSTICE POLICY (1977); Ashlee Willis, *Community Service as an Alternative to Imprisonment: A Cautionary View*, 24 PROBATION J. 120 (1977).

57. Julie Leibrich, Burt Galaway, & Yvonne Underhill, *Community Sentencing in New Zealand: A Survey of Users*, 50 FED. PROBATION 55 (1986).

58. James Austin & Barry Krisberg, *The Unmet Promise of Alternatives*, 28 J. RES. CRIME & DELINQ. 374 (1982); Mark S. Umbreit, *Community Service Sentencing: Jail Alternatives or Added Sanction?*, 45 FED. PROBATION 3 (1981).

59. FIORI RINALDI, IMPRISONMENT FOR NON-PAYMENT OF FINES (1976); GERHARDT GREBING, THE FINE IN COMPARATIVE LAW: A SURVEY OF 21 COUNTRIES (1982).

60. LEON RADZINOWICZ & ROGER HOOD, A HISTORY OF ENGLISH CRIMINAL LAW AND ITS ADMINISTRATION FROM 1750 VOL. 5: THE EMERGENCE OF PENAL POLICY (1986); PETER YOUNG, PUNISHMENT, MONEY AND THE LEGAL ORDER: AN ANALYSIS OF THE EMERGENCE OF MONETARY SANCTIONS WITH SPECIAL REFERENCE TO SCOTLAND (1987).

61. Gail S. Funke, *The Economics of Prison Crowding*, 478 ANNALS AM. ACAD. POL. & SOC. SCI. 86 (1985); Thomas Mathiesen, *The Viewer Society: Michel Foucault's 'Panopticon' Revisited*, 1 THEORETICAL CRIMINOLOGY 215 (1997).

62. SALLY T. HILLSMAN & SILVIA S. G. CASALE, ENFORCEMENT OF FINES AS CRIMINAL SANCTIONS: THE ENGLISH EXPERIENCE AND ITS RELEVANCE TO AMERICAN PRACTICE (1986); Judith A. Greene, *Structuring Criminal Fines: Making an 'Intermediate Penalty' More Useful and Equitable*, 13 JUSTICE SYSTEM J. 37

(1988); NIGEL WALKER & NICOLA PADFIELD, SENTENCING: THEORY, LAW AND PRACTICE (1996).

63. MICHAEL H. TONRY & KATHLEEN HATLESTAD, SENTENCING REFORM IN OVERCROWDED TIMES: A COMPARATIVE PERSPECTIVE (1997).

64. See section 6.2.4.

65. United States v. Allegheny Bottling Company, 695 F.Supp. 856 (1988).

SELECTED BIBLIOGRAPHY

FRANZ ALEXANDER & HUGO STAUB, THE CRIMINAL, THE JUDGE, AND THE PUBLIC (1931)

FRANZ ALEXANDER, OUR AGE OF UNREASON: A STUDY OF THE IRRATIONAL FORCES IN SOCIAL LIFE (rev. ed. 1971)

HARRY E. ALLEN, ERIC W. CARLSON, & EVALYN C. PARKS, CRITICAL ISSUES IN ADULT PROBATION (1979)

Susan M. Allan, *No Code Orders v. Resuscitation: The Decision to Withhold Life-Prolonging Treatment from the Terminally Ill*, 26 WAYNE L. REV. 139 (1980)

Peter Alldridge, *The Doctrine of Innocent Agency*, 2 CRIM. L. F. 45 (1990)

FRANCIS ALLEN, THE DECLINE OF THE REHABILITATIVE IDEAL (1981)

Marc Ancel, *The System of Conditional Sentence or Sursis*, 80 L. Q. REV. 334 (1964)

MARC ANCEL, SUSPENDED SENTENCE (1971)

Johannes Andenaes, *The General Preventive Effects of Punishment*, 114 U. PA. L. REV. 949 (1966)

Johannes Andenaes, *The Morality of Deterrence*, 37 U. CHI. L. REV. 649 (1970)

Susan Leigh Anderson, *Asimov's "Three Laws of Robotics" and Machine Metaethics*, 22 AI SOC. 477 (2008)

Edward B. Arnolds & Norman F. Garland, *The Defense of Necessity in Criminal Law: The Right to Choose the Lesser Evil*, 65 J. CRIM. L. & CRIMINOLOGY 289 (1974)

Andrew Ashworth, *The Scope of Criminal Liability for Omissions*, 84 L. Q. REV. 424 (1989)

Andrew Ashworth, *Testing Fidelity to Legal Values: Official Involvement and Criminal Justice*, 63 MOD. L. REV. 663 (2000)

ANDREW ASHWORTH, PRINCIPLES OF CRIMINAL LAW (5th ed. 2006)

Andrew Ashworth, *Rehabilitation, in* PRINCIPLED SENTENCING: READINGS ON THEORY AND POLICY 1 (Andrew von Hirsch, Andrew Ashworth, & Julian Roberts eds., 3d ed. 2009)

ISAAC ASIMOV, I, ROBOT (1950)

ISSAC ASIMOV, THE REST OF ROBOTS (1964)

Tom Athanasiou, *High-Tech Politics: The Case of Artificial Intelligence*, 92 SOCIALIST REV. 7 (1987)

James Austin & Barry Krisberg, *The Unmet Promise of Alternatives*, 28 J. RES. CRIME & DELINQ. (1982)

JOHN AUSTIN, THE PROVINCE OF JURISPRUDENCE DETERMINED (1832, 2000)

BERNARD BAARS, IN THE THEATRE OF CONSCIOUSNESS (1997)

John S. Baker, Jr., *State Police Powers and the Federalization of Local Crime*, 72 TEMP. L. REV. 673 (1999)

STEPHEN BAKER, FINAL JEOPARDY: MAN VS. MACHINE AND THE QUEST TO KNOW EVERYTHING (2011)

CESARE BECCARIA, TRAITÉ DES DÉLITS ET DES PEINES (1764)

Hugo Adam Bedau, *Abolishing the Death Penalty Even for the Worst Murderers*, in THE KILLING STATE — CAPITAL PUNISHMENT IN LAW, POLITICS, AND CULTURE 40 (Austin Sarat ed., 1999)

RICHARD E. BELLMAN, AN INTRODUCTION TO ARTIFICIAL INTELLIGENCE: CAN COMPUTERS THINK? (1978)

JEREMY BENTHAM, AN INTRODUCTION TO THE PRINCIPLES OF MORALS AND LEGISLATION (1789, 1996)

Jeremy Bentham, *Punishment and Deterrence*, in PRINCIPLED SENTENCING: READINGS ON THEORY AND POLICY (Andrew von Hirsch, Andrew Ashworth, & Julian Roberts eds., 3d ed. 2009)

JOHN BIGGS, THE GUILTY MIND (1955)

Ned Block, *What Intuitions About Homunculi Don't Show*, 3 BEHAVIORAL & BRAIN SCI. 425 (1980)

ROBERT M. BOHM, DEATHQUEST: AN INTRODUCTION TO THE THEORY AND PRACTICE OF CAPITAL PUNISHMENT IN THE UNITED STATES (1999)

Addison M. Bowman, *Narcotic Addiction and Criminal Responsibility under Durham*, 53 GEO. L.J. 1017 (1965)

STEVEN BOX, POWER, CRIME AND MYSTIFICATION (1983)

RICHARD B. BRANDT, ETHICAL THEORY (1959)

Kathleen F. Brickey, *Corporate Criminal Accountability: A Brief History and an Observation*, 60 WASH. U. L. Q. 393 (1983)

Bruce Bridgeman, *Brains + Programs = Minds*, 3 BEHAVIORAL & BRAIN SCI. 427 (1980)

WALTER BROMBERG, FROM SHAMAN TO PSYCHOTHERAPIST: A HISTORY OF THE TREATMENT OF MENTAL ILLNESS (1975)

Andrew G. Brooks & Ronald C. Arkin, *Behavioral Overlays for Non-Verbal Communication Expression on a Humanoid Robot*, 22 AUTON. ROBOTS 55 (2007)

Timothy L. Butler, *Can a Computer Be an Author—Copyright Aspects of Artificial Intelligence*, 4 COMM. ENT. L. S. 707 (1982)

Kenneth L. Campbell, *Psychological Blow Automatism: A Narrow Defence*, 23 CRIM. L. Q. 342 (1981)

W. G. Carson, *Some Sociological Aspects of Strict Liability and the Enforcement of Factory Legislation*, 33 MOD. L. REV. 396 (1970)

W. G. Carson, *The Conventionalisation of Early Factory Crime*, 7 INT'L J. OF SOCIOLOGY OF LAW 37 (1979)

Derrick Augustus Carter, *Bifurcations of Consciousness: The Elimination of the Self-Induced Intoxication Excuse*, 64 MO. L. REV. 383 (1999)

MICHAEL CAVADINO & JAMES DIGNAN, THE PENAL SYSTEM: AN INTRODUCTION (2002)

EUGENE CHARNIAK & DREW MCDERMOTT, INTRODUCTION TO ARTIFICIAL INTELLIGENCE (1985)

Russell L. Christopher, *Deterring Retributivism: The Injustice of "Just" Punishment*, 96 NW. U. L. REV. 843 (2002)

ARTHUR C. CLARKE, 2001: A SPACE ODYSSEY (1968)

John C. Coffee, Jr., *"No Soul to Damn: No Body to Kick": An Unscandalised Inquiry into the Problem of Corporate Punishment*, 79 MICH. L. REV. 386 (1981)

SIR EDWARD COKE, INSTITUTIONS OF THE LAWS OF ENGLAND — THIRD PART (6th ed. 1681, 1817, 2001)

Dana K. Cole, *Expending Felony-Murder in Ohio: Felony-Murder or Murder-Felony*, 63 OHIO ST. L.J. 15 (2002)

ROBERTA C. CRONIN, BOOT CAMPS FOR ADULT AND JUVENILE OFFENDERS: OVERVIEW AND UPDATE (1994)

George R. Cross & Cary G. Debessonet, *An Artificial Intelligence Application in the Law: CCLIPS, A Computer Program that Processes Legal Information*, 1 HIGH TECH. L.J. 329 (1986)

Homer D. Crotty, *The History of Insanity as a Defence to Crime in English Common Law*, 12 CAL. L. REV. 105 (1924)

MICHAEL DALTON, THE COUNTREY JUSTICE (1618, 2003)

Donald Davidson, *Turing's Test*, in MODELLING THE MIND (K. A. Mohyeldin Said, W. H. Newton-Smith, R. Viale, & K. V. Wilkes eds., 1990)

Michael J. Davidson, *Feminine Hormonal Defenses: Premenstrual Syndrome and Postpartum Psychosis*, 5 ARMY LAWYER (2000)

Richard Delgado, *Ascription of Criminal States of Mind: Toward a Defense Theory for the Coercively Persuaded ("Brainwashed") Defendant*, 63 MINN. L. REV. 1 (1978)

DANIEL C. DENNETT, BRAINSTORMS (1978)

DANIEL C. DENNETT, THE INTENTIONAL STANCE (1987)

Daniel C. Dennett, *Evolution, Error, and Intentionality*, in THE FOUNDATIONS OF ARTIFICIAL INTELLIGENCE 190 (Derek Pertridge & Yorick Wilks eds., 1990, 2006)

RENÉ DESCARTES, DISCOURS DE LA MÉTHODE POUR BIEN CONDUIRE SA RAISON ET CHERCHER LA VÉRITÉ DANS LES SCIENCES (1637)

Anthony M. Dillof, *Unraveling Unknowing Justification*, 77 NOTRE DAME L. REV. 1547 (2002)

Dolores A. Donovan & Stephanie M. Wildman, *Is the Reasonable Man Obsolete? A Critical Perspective on Self-Defense and Provocation*, 14 LOY. L. A. L. REV. 435, 441 (1981)

AAGE GERHARDT DRACHMANN, THE MECHANICAL TECHNOLOGY OF GREEK AND ROMAN ANTIQUITY: A STUDY OF THE LITERARY SOURCES (1963)

Joshua Dressler, *Professor Delgado's "Brainwashing" Defense: Courting a Determinist Legal System*, 63 MINN. L. REV. 335 (1978)

Joshua Dressler, *Rethinking Heat of Passion: A Defense in Search of a Rationale*, 73 J. CRIM. L. & CRIMINOLOGY 421 (1982)

Joshua Dressler, *Battered Women Who Kill Their Sleeping Tormenters: Reflections on Maintaining Respect for Human Life while Killing Moral Monsters*, in CRIMINAL LAW THEORY — DOCTRINES OF THE GENERAL PART 259 (Stephen Shute & A. P. Simester eds., 2005)

G. R. DRIVER & JOHN C. MILES, THE BABYLONIAN LAWS, VOL. I: LEGAL COMMENTARY (1952)

ANTONY ROBIN DUFF, CRIMINAL ATTEMPTS (1996)

Fernand N. Dutile & Harold F. Moore, *Mistake and Impossibility: Arranging Marriage Between Two Difficult Partners*, 74 NW. U. L. REV. 166 (1980)

Justice Ellis, *Criminal Law as an Instrument of Social Control*, 17 VICTORIA U. WELLINGTON L. REV. 319 (1987)

GERTRUDE EZORSKY, PHILOSOPHICAL PERSPECTIVES ON PUNISHMENT (1972)

Judith Fabricant, *Homicide in Response to a Threat of Rape: A Theoretical Examination of the Rule of Justification*, 11 GOLDEN GATE U. L. REV. 945 (1981)

DAVID P. FARRINGTON & BRANDON C. WELSH, PREVENTING CRIME: WHAT WORKS FOR CHILDREN, OFFENDERS, VICTIMS AND PLACES (2006)

EDWARD A. FEIGENBAUM & PAMELA MCCORDUCK, THE FIFTH GENERATION: ARTIFICIAL INTELLIGENCE AND JAPAN'S COMPUTER CHALLENGE TO THE WORLD (1983)

S. Z. Feller, *Les Délits de Mise en Danger*, 40 REV. INT. DE DROIT PÉNAL 179 (1969)

JAMIE FELLNER & JOANNE MARINER, COLD STORAGE: SUPER-MAXIMUM SECURITY CONFINEMENT IN INDIANA (1997)

Robert P. Fine & Gary M. Cohen, *Is Criminal Negligence a Defensible Basis for Criminal Liability?*, 16 BUFF. L. REV. 749 (1966)

PAUL JOHANN ANSELM FEUERBACH, LEHRBUCH DES GEMEINEN IN DEUTSCHLAND GÜLTIGEN PEINLICHEN RECHTS (1812, 2007)

Stuart Field & Nico Jorg, *Corporate Liability and Manslaughter: Should We Be Going Dutch?*, [1991] CRIM. L.R. 156 (1991)

Herbert Fingarette, *Addiction and Criminal Responsibility*, 84 YALE L.J. 413 (1975)

ARTHUR E. FINK, CAUSES OF CRIME: BIOLOGICAL THEORIES IN THE UNITED STATES, 1800–1915 (1938)

JOHN FINNIS, NATURAL LAW AND NATURAL RIGHTS (1980)

Brent Fisse & John Braithwaite, *The Allocation of Responsibility for Corporate Crime: Individualism, Collectivism and Accountability*, 11 SYDNEY L. REV. 468 (1988)

Peter Fitzpatrick, *"Always More to Do": Capital Punishment and the (De)Composition of Law*, in THE KILLING STATE — CAPITAL PUNISHMENT IN LAW, POLITICS, AND CULTURE 117 (Austin Sarat ed., 1999)

OWEN J. FLANAGAN, JR., THE SCIENCE OF THE MIND (2d ed. 1991)
GEORGE P. FLETCHER, RETHINKING CRIMINAL LAW (1978, 2000)
George P. Fletcher, *The Nature of Justification, in* ACTION AND VALUE IN CRIMINAL LAW 175 (Stephen Shute, John Gardner, & Jeremy Horder eds., 2003)
Jerry A. Fodor, *Modules, Frames, Fridgeons, Sleeping Dogs and the Music of the Spheres, in* THE ROBOT'S DILEMMA: THE FRAME PROBLEM IN ARTIFICIAL INTELLIGENCE (Zenon W. Pylyshyn ed., 1987)
Keith Foren, *Casenote: In Re Tyvonne M. Revisited: The Criminal Infancy Defense in Connecticut*, 18 Q. L. REV. 733 (1999)
MICHEL FOUCAULT, MADNESS AND CIVILIZATION (1965)
MICHEL FOUCAULT, DISCIPLINE AND PUNISH: THE BIRTH OF THE PRISON (1977)
Sue Frank, *Oklahoma Camp Stresses Structure and Discipline*, 53 CORRECTIONS TODAY 102 (1991)
Lionel H. Frankel, *Criminal Omissions: A Legal Microcosm*, 11 WAYNE L. REV. 367 (1965)
Lionel H. Frankel, *Narcotic Addiction, Criminal Responsibility and Civil Commitment*, 1966 UTAH L. REV. 581 (1966)
Robert M. French, *Subcognition and the Limits of the Turing Test*, 99 MIND 53 (1990)
K. W. M. Fulford, *Value, Action, Mental Illness, and the Law, in* ACTION AND VALUE IN CRIMINAL LAW 279 (Stephen Shute, John Gardner, & Jeremy Horder eds., 2003)
Gail S. Funke, *The Economics of Prison Crowding*, 478 ANNALS AM. ACAD. POL. & SOC. SCI. 86 (1985)
Jonathan M. E. Gabbai, Complexity and the Aerospace Industry: Understanding Emergence by Relating Structure to Performance Using Multi-Agent Systems (PhD Thesis, University of Manchester, 2005)
Crystal A. Garcia, *Using Palmer's Global Approach to Evaluate Intensive Supervision Programs: Implications for Practice*, 4 CORRECTION MANAG. Q. 60 (2000)
HOWARD GARDNER, THE MIND'S NEW SCIENCE: A HISTORY OF THE COGNITIVE REVOLUTION (1985)
DAVID GARLAND, THE CULTURE OF CONTROL: CRIME AND SOCIAL ORDER IN CONTEMPORARY SOCIETY (2002)
Chas E. George, *Limitation of Police Powers*, 12 LAW. & BANKER & S. BENCH & B. REV. 740 (1919)
Jack P. Gibbs, *A Very Short Step toward a General Theory of Social Control*, 1985 AM. B. FOUND RES. J. 607 (1985)
SANDER L. GILMAN, SEEING THE INSANE (1982)
P. R. Glazebrook, *Criminal Omissions: The Duty Requirement in Offences Against the Person*, 55 L. Q. REV. 386 (1960)
ROBERT M. GLORIOSO & FERNANDO C. COLON OSORIO, ENGINEERING INTELLIGENT SYSTEMS: CONCEPTS AND APPLICATIONS (1980)
Sheldon Glueck, *Principles of a Rational Penal Code*, 41 HARV. L. REV. 453 (1928)

SIR GERALD GORDON, THE CRIMINAL LAW OF SCOTLAND (1st ed. 1967)
GERHARDT GREBING, THE FINE IN COMPARATIVE LAW: A SURVEY OF 21 COUNTRIES (1982)
Kent Greenawalt, *The Perplexing Borders of Justification and Excuse*, 84 COLUM. L. REV. 1897 (1984)
Kent Greenawalt, *Distinguishing Justifications from Excuses*, 49 LAW & CONTEMP. PROBS. 89 (1986)
David F. Greenberg, *The Corrective Effects of Corrections: A Survey of Evaluation*, in CORRECTIONS AND PUNISHMENT 111 (David F. Greenberg ed., 1977)
Judith A. Greene, *Structuring Criminal Fines: Making an 'Intermediate Penalty' More Useful and Equitable*, 13 JUSTICE SYSTEM J. 37 (1988)
Richard Gruner, *To Let the Punishment Fit the Organization: Sanctioning Corporate Offenders Through Corporate Probation*, 16 AM. J. CRIM. L. 1 (1988)
JEROME HALL, GENERAL PRINCIPLES OF CRIMINAL LAW (2d ed. 1960, 2005)
Jerome Hall, *Intoxication and Criminal Responsibility*, 57 HARV. L. REV. 1045 (1944)
Jerome Hall, *Negligent Behaviour Should Be Excluded from Penal Liability*, 63 COLUM. L. REV. 632 (1963)
Seymour L. Halleck, *The Historical and Ethical Antecedents of Psychiatric Criminology*, in PSYCHIATRIC ASPECTS OF CRIMINOLOGY 8 (Halleck & Bromberg eds., 1968)
Gabriel Hallevy, *The Recidivist Wants to Be Punished—Punishment as an Incentive to Re-offend*, 5 INT'L J. PUNISHMENT & SENTENCING 124 (2009)
GABRIEL HALLEVY, A MODERN TREATISE ON THE PRINCIPLE OF LEGALITY IN CRIMINAL LAW (2010)
Gabriel Hallevy, *The Criminal Liability of Artificial Intelligence Entities—from Science Fiction to Legal Social Control*, 4 AKRON INTELL. PROP. J. 171 (2010)
Gabriel Hallevy, *Unmanned Vehicles—Subordination to Criminal Law under the Modern Concept of Criminal Liability*, 21 J. L. INF. & SCI. 200 (2011)
Gabriel Hallevy, *Therapeutic Victim-Offender Mediation within the Criminal Justice Process—Sharpening the Evaluation of Personal Potential for Rehabilitation while Righting Wrongs under the Alternative-Dispute-Resolution (ADR) Philosophy*, 16 HARV. NEGOT. L. REV. 65 (2011)
GABRIEL HALLEVY, THE MATRIX OF DERIVATIVE CRIMINAL LIABILITY (2012)
John Harding, *The Development of the Community Service*, in ALTERNATIVE STRATEGIES FOR COPING WITH CRIME 164 (Norman Tutt ed., 1978)
HERBERT L. A. HART, PUNISHMENT AND RESPONSIBILITY: ESSAYS IN THE PHILOSOPHY OF LAW (1968)
Frank E. Hartung, *Trends in the Use of Capital Punishment*, 284(1) ANNALS AM. ACAD. POL. & SOC. SCI. 8 (1952)
John Haugeland, *Semantic Engines: An Introduction to Mind Design*, in MIND DESIGN 1 (John Haugeland ed., 1981)

JOHN HAUGELAND, ARTIFICIAL INTELLIGENCE: THE VERY IDEA (1985)
PAMELA RAE HEATH, THE PK ZONE: A CROSS-CULTURAL REVIEW OF PSYCHOKINESIS (PK) (2003)
HERMANN VON HELMHOLTZ, THE FACTS OF PERCEPTION (1878)
John Lawrence Hill, *A Utilitarian Theory of Duress*, 84 IOWA L. REV. 275 (1999)
SALLY T. HILLSMAN & SILVIA S. G. CASALE, ENFORCEMENT OF FINES AS CRIMINAL SANCTIONS: THE ENGLISH EXPERIENCE AND ITS RELEVANCE TO AMERICAN PRACTICE (1986)
Harold L. Hirsh & Richard E. Donovan, *The Right to Die: Medico-Legal Implications of In re Quinlan*, 30 RUTGERS L. REV. 267 (1977)
ANDREW VON HIRSCH, DOING JUSTICE: THE CHOICE OF PUNISHMENT (1976)
Andrew von Hirsch, *Proportionate Sentences: A Desert Perspective*, in PRINCIPLED SENTENCING: READINGS ON THEORY AND POLICY 115 (Andrew von Hirsch, Andrew Ashworth, & Julian Roberts eds., 3d ed. 2009)
W. H. Hitchler, *Necessity as a Defence in Criminal Cases*, 33 DICK. L. REV. 138 (1929)
THOMAS HOBBES, LEVIATHAN OR THE MATTER, FORME AND POWER OF A COMMON WEALTH ECCLESIASTICALL AND CIVIL (1651)
DOUGLAS R. HOFSTADTER, GÖDEL, ESCHER, BACH: AN ETERNAL GOLDEN BRAID (1979, 1999)
William Searle Holdsworth, *English Corporation Law in the 16th and 17th Centuries*, 31 YALE L.J. 382 (1922)
WILLIAM SEARLE HOLDSWORTH, A HISTORY OF ENGLISH LAW (1923)
Winifred H. Holland, *Automatism and Criminal Responsibility*, 25 CRIM. L. Q. 95 (1982)
Clive R. Hollin, *Treatment Programs for Offenders*, 22 INT'L J. OF LAW & PSYCHIATRY 361 (1999)
OLIVER W. HOLMES, THE COMMON LAW (1881, 1923)
Oliver W. Holmes, *Agency*, 4 HARV. L. REV. 345 (1891)
HENRY HOLT, TELEKINESIS (2005)
Morton J. Horwitz, *The Rise and Early Progressive Critique of Objective Causation*, in THE POLITICS OF LAW: A PROGRESSIVE CRITIQUE 471 (David Kairys ed., 3d ed. 1998)
JOHN HOWARD, THE STATE OF PRISONS IN ENGLAND AND WALES (1777, 1996)
FENG-HSIUNG HSU, BEHIND DEEP BLUE: BUILDING THE COMPUTER THAT DEFEATED THE WORLD CHESS CHAMPION (2002)
BARBARA HUDSON, UNDERSTANDING JUSTICE: AN INTRODUCTION TO IDEAS, PERSPECTIVES AND CONTROVERSIES IN MODERN PENAL THEORY (1996, 2003)
Graham Hughes, *Criminal Omissions*, 67 YALE L.J. 590 (1958)
BISHOP CARLETON HUNT, THE DEVELOPMENT OF THE BUSINESS CORPORATION IN ENGLAND 1800–1867 (1963)
Douglas Husak, *Holistic Retribution*, 88 CAL. L. REV. 991 (2000)
Douglas Husak, *Retribution in Criminal Theory*, 37 SAN DIEGO L. REV. 959 (2000)

Douglas Husak & Andrew von Hirsch, *Culpability and Mistake of Law*, in ACTION AND VALUE IN CRIMINAL LAW 157 (Stephen Shute, John Gardner, & Jeremy Horder eds., 2003)

Peter Barton Hutt & Richard A. Merrill, *Criminal Responsibility and the Right to Treatment for Intoxication and Alcoholism*, 57 GEO. L.J. 835 (1969)

RAY JACKENDOFF, CONSCIOUSNESS AND THE COMPUTATIONAL MIND (1987)

WILLIAM JAMES, THE PRINCIPLES OF PSYCHOLOGY (1890)

PHILLIP N. JOHNSON-LAIRD, MENTAL MODELS (1983)

Matthew Jones, *Overcoming the Myth of Free Will in Criminal Law: The True Impact of the Genetic Revolution*, 52 DUKE L.J. 1031 (2003)

Sanford Kadish, *Respect for Life and Regard for Rights in the Criminal Law*, 64 CAL. L. REV. 871 (1976)

Sanford H. Kadish, *Excusing Crime*, 75 CAL. L. REV. 257 (1987)

Martin P. Kafka, *Sex Offending and Sexual Appetite: The Clinical and Theoretical Relevance of Hypersexual Desire*, 47 INT'L J. OF OFFENDER THERAPY & COMP. CRIMINOLOGY 439 (2003)

IMMANUEL KANT, OUR DUTIES TO ANIMALS (1780)

A. W. G. Kean, *The History of the Criminal Liability of Children*, 53 L. Q. REV. 364 (1937)

VOJISLAV KECMAN, LEARNING AND SOFT COMPUTING, SUPPORT VECTOR MACHINES, NEURAL NETWORKS AND FUZZY LOGIC MODELS (2001)

Edwin R. Keedy, *Ignorance and Mistake in the Criminal Law*, 22 HARV. L. REV. 75 (1909)

Paul W. Keve, *The Professional Character of the Presentence Report*, 26 FED. PROBATION 51 (1962)

ANTONY KENNY, WILL, FREEDOM AND POWER (1975)

ANTONY KENNY, WHAT IS FAITH? (1992)

Roy D. King, *The Rise and Rise of Supermax: An American Solution in Search of a Problem?*, 1 PUNISHMENT & SOCIETY 163 (1999)

RAYMOND KURZWEIL, THE AGE OF INTELLIGENT MACHINES (1990)

NICOLA LACEY & CELIA WELLS, RECONSTRUCTING CRIMINAL LAW — CRITICAL PERSPECTIVES ON CRIME AND THE CRIMINAL PROCESS (2d ed. 1998)

NICOLA LACEY, CELIA WELLS, & OLIVER QUICK, RECONSTRUCTING CRIMINAL LAW (3d ed. 2003, 2006)

J. G. LANDELS, ENGINEERING IN THE ANCIENT WORLD (rev. ed. 2000)

William S. Laufer, *Corporate Bodies and Guilty Minds*, 43 EMORY L.J. 647 (1994)

GOTTFRIED WILHELM LEIBNIZ, CHARACTERISTICA UNIVERSALIS (1676)

Julie Leibrich, Burt Galaway, & Yvonne Underhill, *Community Sentencing in New Zealand: A Survey of Users*, 50 FED. PROBATION 55 (1986)

LAWRENCE LESSIG, CODE AND OTHER LAWS OF CYBERSPACE (1999)

DAVID LEVY, ROBOTS UNLIMITED: LIFE IN A VIRTUAL AGE (2006)

DAVID LEVY, LOVE AND SEX WITH ROBOTS: THE EVOLUTION OF HUMAN-ROBOT RELATIONSHIPS (2007)

DAVID LEVY & MONTY NEWBORN, HOW COMPUTERS PLAY CHESS (1991)

K. W. Lidstone, *Social Control and the Criminal Law*, 27 BRIT. J. CRIMINOLOGY 31 (1987)

DOUGLAS S. LIPTON, ROBERT MARTINSON, & JUDITH WILKS, THE EFFECTIVENESS OF CORRECTIONAL TREATMENT: A SURVEY OF TREATMENT EVALUATION STUDIES (1975)

Frederick J. Ludwig, *Rationale of Responsibility for Young Offenders*, 29 NEB. L. REV. 521 (1950)

GEORGE F. LUGER, ARTIFICIAL INTELLIGENCE: STRUCTURES AND STRATEGIES FOR COMPLEX PROBLEM SOLVING (2001)

GEORGE F. LUGER & WILLIAM A. STUBBLEFIELD, ARTIFICIAL INTELLIGENCE: STRUCTURES AND STRATEGIES FOR COMPLEX PROBLEM SOLVING (6th ed. 2008)

William G. Lycan, *Introduction, in* MIND AND COGNITION 3 (William G. Lycan ed., 1990)

Gerard E. Lynch, *The Role of Criminal Law in Policing Corporate Misconduct*, 60 LAW & CONTEMP. PROBS. 23 (1997)

Peter Lynch, *The Origins of Computer Weather Prediction and Climate Modeling*, 227 JOURNAL OF COMPUTATIONAL PHYSICS 3431 (2008)

DAVID LYONS, FORMS AND LIMITS OF UTILITARIANISM (1965)

David Lyons, *Open Texture and the Possibility of Legal Interpretation*, 18 LAW PHIL. 297 (1999)

DORRIS LAYTON MACKANZIE & EUGENE E. HEBERT, CORRECTIONAL BOOT CAMPS: A TOUGH INTERMEDIATE SANCTION (1996)

BRONISLAW MALINOWSKI, CRIME AND CUSTOM IN SAVAGE SOCIETY (1959, 1982)

DAVID MANNERS & TSUGIO MAKIMOTO, LIVING WITH THE CHIP (1995)

Dan Markel, *Are Shaming Punishments Beautifully Retributive? Retributivism and the Implications for the Alternative Sanctions Debate*, 54 VAND. L. REV. 2157 (2001)

Robert Martinson, *What Works? Questions and Answers about Prison Reform*, 35 PUBLIC INTEREST 22 (1974)

Thomas Mathiesen, *The Viewer Society: Michel Foucault's 'Panopticon' Revisited*, 1 THEORETICAL CRIMINOLOGY 215 (1997)

Peter McCandless, *Liberty and Lunacy: The Victorians and Wrongful Confinement, in* MADHOUSES, MAD-DOCTORS, AND MADMEN: THE SOCIAL HISTORY OF PSYCHIATRY IN THE VICTORIAN ERA (Scull ed., 1981)

Aileen McColgan, *In Defence of Battered Women who Kill*, 13 OXFORD J. LEGAL STUD. 508 (1993)

Sean McConville, *The Victorian Prison: England 1865–1965, in* THE OXFORD HISTORY OF THE PRISON 131 (Norval Morris & David J. Rothman eds., 1995)

J. R. MCDONALD, G. M. BURT, J. S. ZIELINSKI, & S. D. J. MCARTHUR, INTELLIGENT KNOWLEDGE BASED SYSTEM IN ELECTRICAL POWER ENGINEERING (1997)

COLIN MCGINN, THE PROBLEM OF CONSCIOUSNESS: ESSAYS TOWARDS A RESOLUTION (1991)

KARL MENNINGER, MARTIN MAYMAN, & PAUL PRUYSER, THE VITAL BALANCE (1963)

Alan C. Michaels, *Imposing Constitutional Limits on Strict Liability: Lessons from the American Experience*, in APPRAISING STRICT LIABILITY 218 (A. P. Simester ed., 2005)

DONALD MICHIE & RORY JOHNSTON, THE CREATIVE COMPUTER (1984)

Justine Miller, *Criminal Law—An Agency for Social Control*, 43 YALE L.J. 691 (1934)

MARVIN MINSKY, THE SOCIETY OF MIND (1986)

JESSICA MITFORD, KIND AND USUAL PUNISHMENT: THE PRISON BUSINESS (1974)

Patrick Montague, *Self-Defense and Choosing Between Lives*, 40 PHIL. STUD. 207 (1981)

MICHAEL MOORE, LAW AND PSYCHIATRY: RETHINKING THE RELATIONSHIP (1984)

George Mora, *Historical and Theoretical Trends in Psychiatry*, in COMPREHENSIVE TEXTBOOK OF PSYCHIATRY 1 (Alfred M. Freedman, Harold Kaplan, & Benjamin J. Sadock eds., 2d ed. 1975)

HANS MORAVEC, ROBOT: MERE MACHINE TO TRANSCENDENT MIND (1999)

Norval Morris, *Somnambulistic Homicide: Ghosts, Spiders, and North Koreans*, 5 RES JUDICATAE 29 (1951)

TIM MORRIS, COMPUTER VISION AND IMAGE PROCESSING (2004)

Gerhard O. W. Mueller, *Mens Rea and the Corporation—A Study of the Model Penal Code Position on Corporate Criminal Liability*, 19 U. PITT. L. REV. 21 (1957)

Michael A. Musmanno, *Are Subordinate Officials Penally Responsible for Obeying Superior Orders which Direct Commission of Crime?*, 67 DICK. L. REV. 221 (1963)

MONTY NEWBORN, DEEP BLUE (2002)

ALLEN NEWELL & HERBERT A. SIMON, HUMAN PROBLEM SOLVING (1972)

EDWARD NORBECK, RELIGION IN PRIMITIVE SOCIETY (1961)

Anne Norton, *After the Terror: Mortality, Equality, Fraternity*, in THE KILLING STATE—CAPITAL PUNISHMENT IN LAW, POLITICS, AND CULTURE 27 (Austin Sarat ed., 1999)

Scott T. Noth, *A Penny for Your Thoughts: Post-Mitchell Hate Crime Laws Confirm a Mutating Effect upon Our First Amendment and the Government's Role in Our Lives*, 10 REGENT U. L. REV. 167 (1998)

DAVID ORMEROD, SMITH & HOGAN CRIMINAL LAW (11th ed. 2005)

N. P. PADHY, ARTIFICIAL INTELLIGENCE AND INTELLIGENT SYSTEMS (2005, 2009)

WILLIAM PALEY, A TREATISE ON THE LAW OF PRINCIPAL AND AGENT (2d ed. 1847)

DAN W. PATTERSON, INTRODUCTION TO ARTIFICIAL INTELLIGENCE AND EXPERT SYSTEMS (1990)

Monrad G. Paulsen, *Intoxication as a Defense to Crime*, 1961 U. ILL. L. F. 1 (1961)

Rollin M. Perkins, *Negative Acts in Criminal Law*, 22 IOWA L. REV. 659 (1937)

Rollin M. Perkins, *Ignorance and Mistake in Criminal Law*, 88 U. PA. L. REV. 35 (1940)

Rollin M. Perkins, *"Knowledge" as a Mens Rea Requirement*, 29 HASTINGS L.J. 953 (1978)

Rollin M. Perkins, *Impelled Perpetration Restated*, 33 HASTINGS L.J. 403 (1981)

ANTHONY M. PLATT, THE CHILD SAVERS: THE INVENTION OF DELINQUENCY (2d ed. 1969, 1977)

Anthony Platt & Bernard L. Diamond, *The Origins of the "Right and Wrong" Test of Criminal Responsibility and Its Subsequent Development in the United States: An Historical Survey*, 54 CAL. L. REV. 1227 (1966)

FREDERICK POLLOCK & FREDERICK WILLIAM MAITLAND, THE HISTORY OF ENGLISH LAW BEFORE THE TIME OF EDWARD I (rev. 2d ed. 1898)

Stanislaw Pomorski, *On Multiculturalism, Concepts of Crime, and the "De Minimis" Defense*, 1997 B.Y.U. L. REV. 51 (1997)

JAMES COWLES PRICHARD, A TREATISE ON INSANITY AND OTHER DISORDERS AFFECTING THE MIND (1835)

GUSTAV RADBRUCH, DER HANDLUNGSBEGRIFF IN SEINER BEDEUTUNG FÜR DAS STRAFRECHTSSYSTEM (1904)

LEON RADZINOWICZ, A HISTORY OF ENGLISH CRIMINAL LAW AND ITS ADMINISTRATION FROM 1750 VOL. 1: THE MOVEMENT FOR REFORM (1948)

LEON RADZINOWICZ & ROGER HOOD, A HISTORY OF ENGLISH CRIMINAL LAW AND ITS ADMINISTRATION FROM 1750 VOL. 5: THE EMERGENCE OF PENAL POLICY (1986)

Craig W. Reynolds, *Herds and Schools: A Distributed Behavioral Model*, 21 COMPUT. GRAPH. (1987)

ELAINE RICH & KEVIN KNIGHT, ARTIFICIAL INTELLIGENCE (2d ed. 1991)

FIORI RINALDI, IMPRISONMENT FOR NON-PAYMENT OF FINES (1976)

Edwina L. Rissland, *Artificial Intelligence and Law: Stepping Stones to a Model of Legal Reasoning*, 99 YALE L.J. 1957 (1990)

CHASE RIVELAND, SUPERMAX PRISONS: OVERVIEW AND GENERAL CONSIDERATIONS (1999)

OLIVIA F. ROBINSON, THE CRIMINAL LAW OF ANCIENT ROME (1995)

Paul H. Robinson, *A Theory of Justification: Societal Harm as a Prerequisite for Criminal Liability*, 23 U.C.L.A. L. REV. 266 (1975)

Paul H. Robinson & John M. Darley, *The Utility of Desert*, 91 NW. U. L. REV. 453 (1997)

Paul H. Robinson, *Testing Competing Theories of Justification*, 76 N.C. L. REV. 1095 (1998)

P. ROGERS, LAW ON THE BATTLEFIELD (1996)

Vashon R. Rogers, Jr., *De Minimis Non Curat Lex*, 21 ALBANY L.J. 186 (1880)

GEORGE ROSEN, MADNESS IN SOCIETY: CHAPTERS IN THE HISTORICAL SOCIOLOGY OF MENTAL ILLNESS (1969)

Laurence H. Ross, *Deterrence Regained: The Cheshire Constabulary's "Breathalyser Blitz,"* 6 J. LEGAL STUD. 241 (1977)

DAVID J. ROTHMAN, CONSCIENCE AND CONVENIENCE: THE ASYLUM AND ITS ALTERNATIVES IN PROGRESSIVE AMERICA (1980)

David J. Rothman, *For the Good of All: The Progressive Tradition in Prison Reform*, in HISTORY AND CRIME: IMPLICATIONS FOR CRIMINAL JUSTICE POLICY 271 (James A. Inciardi & Charles E. Faupel eds., 1980)

CLAUS ROXIN, STRAFRECHT — ALLGEMEINER TEIL I (4 Auf. 2006)
STUART J. RUSSELL & PETER NORVIG, ARTIFICIAL INTELLIGENCE: A MODERN APPROACH (2002)
WILLIAM OLDNALL RUSSELL, A TREATISE ON CRIMES AND MISDEMEANORS (1843, 1964)
Cheyney C. Ryan, *Self-Defense, Pacificism, and the Possibility of Killing*, 93 ETHICS 508 (1983)
GILBERT RYLE, THE CONCEPT OF MIND (1954)
Francis Bowes Sayre, *Criminal Responsibility for the Acts of Another*, 43 HARV. L. REV. 689 (1930)
Francis Bowes Sayre, *Mens Rea*, 45 HARV. L. REV. 974 (1932)
Francis Bowes Sayre, *Public Welfare Offenses*, 33 COLUM. L. REV. 55 (1933).
ROBERT J. SCHALKOFF, ARTIFICIAL INTELLIGENCE: AN ENGINEERING APPROACH (1990)
Roger C. Schank, *What is AI, Anyway?*, in THE FOUNDATIONS OF ARTIFICIAL INTELLIGENCE 3 (Derek Pertridge & Yorick Wilks eds., 1990, 2006)
Samuel Scheffler, *Justice and Desert in Liberal Theory*, 88 CAL. L. REV. 965 (2000)
G. Schoenfeld, *In Defence of Retribution in the Law*, 35 PSYCHOANALYTIC Q. 108 (1966)
FRANK SCHMALLEGER, CRIMINAL JUSTICE TODAY: AN INTRODUCTORY TEXT FOR THE 21ST CENTURY (2003)
WILLIAM ROBERT SCOTT, THE CONSTITUTION AND FINANCE OF ENGLISH, SCOTTISH AND IRISH JOINT-STOCK COMPANIES TO 1720 (1912)
John R. Searle, *Minds, Brains & Programs*, 3 BEHAVIORAL & BRAIN SCI. 417 (1980)
JOHN R. SEARLE, MINDS, BRAINS AND SCIENCE (1984)
JOHN R. SEARLE, THE REDISCOVERY OF MIND (1992)
LEE SECHREST, SUSAN O. WHITE, & ELIZABETH D. BROWN, THE REHABILITATION OF CRIMINAL OFFENDERS: PROBLEMS AND PROSPECTS (1979)
Richard P. Seiter & Karen R. Kadela, *Prisoner Reentry: What Works, What Does Not, and What Is Promising*, 49 CRIME & DELINQUENCY 360 (2003)
THORSTEN J. SELLIN, SLAVERY AND THE PENAL SYSTEM (1976)
Robert N. Shapiro, *Of Robots, Persons, and the Protection of Religious Beliefs*, 56 S. CAL. L. REV. 1277 (1983)
ROSEMARY SHEEHAN, GILL MCLVOR, & CHRIS TROTTER, WHAT WORKS WITH WOMEN OFFENDERS (2007)
LAWRENCE W. SHERMAN, DAVID P. FARRINGTON, DORIS LEYTON MACKENZIE, & BRANDON C. WELSH, EVIDENCE-BASED CRIME PREVENTION (2006)
Nancy Sherman, *The Place of the Emotions in Kantian Morality*, in IDENTITY, CHARACTER, AND MORALITY 145 (Owen Flanagan & Amelie O. Rotry eds., 1990)
Stephen Shute, *Knowledge and Belief in the Criminal Law*, in CRIMINAL LAW THEORY — DOCTRINES OF THE GENERAL PART 182 (Stephen Shute & A.P. Simester eds., 2005)

R. U. Singh, *History of the Defence of Drunkenness in English Criminal Law*, 49 LAW Q. REV. 528 (1933)

VIEDA SKULTANS, ENGLISH MADNESS: IDEAS ON INSANITY, 1580–1890 (1979)

Aaron Sloman, *Motives, Mechanisms, and Emotions, in* THE PHILOSOPHY OF ARTIFICIAL INTELLIGENCE 231 (Margaret A. Boden ed., 1990)

JOHN J. C. SMART & BERNARD WILLIAMS, UTILITARIANISM — FOR AND AGAINST (1973)

RUDOLPH SOHM, THE INSTITUTES OF ROMAN LAW (3d ed. 1907)

Lawrence B. Solum, *Legal Personhood for Artificial Intelligences*, 70 N.C. L. REV. 1231 (1992)

MILAN SONKA, VACLAV HLAVAC, & ROGER BOYLE, IMAGE PROCESSING, ANALYSIS, AND MACHINE VISION (2008)

WALTER W. SOROKA, ANALOG METHODS IN COMPUTATION AND SIMULATION (1954)

John R. Spencer & Antje Pedain, *Approaches to Strict and Constructive Liability in Continental Criminal Law, in* APPRAISING STRICT LIABILITY 237 (A. P. Simester ed., 2005)

Jane Stapelton, *Law, Causation and Common Sense*, 8 OXFORD J. LEGAL STUD. 111 (1988)

G. R. Sullivan, *Knowledge, Belief, and Culpability, in* CRIMINAL LAW THEORY — DOCTRINES OF THE GENERAL PART 207 (Stephen Shute & A. P. Simester eds., 2005)

G. R. Sullivan, *Strict Liability for Criminal Offences in England and Wales Following Incorporation into English Law of the European Convention on Human Rights, in* APPRAISING STRICT LIABILITY 195 (A. P. Simester ed., 2005)

ROGER J. SULLIVAN, IMMANUEL KANT'S MORAL THEORY (1989)

Victor Tadors, *Recklessness and the Duty to Take Care, in* CRIMINAL LAW THEORY — DOCTRINES OF THE GENERAL PART 227 (Stephen Shute & A. P. Simester eds., 2005)

STEVEN L. TANIMOTO, ELEMENTS OF ARTIFICIAL INTELLIGENCE: AN INTRODUCTION USING LISP (1987)

Lawrence Taylor & Katharina Dalton, *Premenstrual Syndrome: A New Criminal Defense?*, 19 CAL. W. L. REV. 269 (1983)

JUDITH JARVIS THOMSON, RIGHTS, RESTITUTION AND RISK: ESSAYS IN MORAL THEORY (1986)

BENJAMIN THORPE, ANCIENT LAWS AND INSTITUTES OF ENGLAND (1840, 2004)

Lawrence P. Tiffany & Carl A. Anderson, *Legislating the Necessity Defense in Criminal Law*, 52 DENV. L.J. 839 (1975)

Janet A. Tighe, *Francis Wharton and the Nineteenth Century Insanity Defense: The Origins of a Reform Tradition*, 27 AM. J. LEGAL HIST. 223 (1983)

Jackson Toby, *Is Punishment Necessary?*, 55 J. CRIM. L. CRIMINOLOGY & POLICE SCI. 332 (1964)

MICHAEL H. TONRY & KATHLEEN HATLESTAD, SENTENCING REFORM IN OVERCROWDED TIMES: A COMPARATIVE PERSPECTIVE (1997)

Richard H. S. Tur, *Subjectivism and Objectivism: Towards Synthesis*, in ACTION AND VALUE IN CRIMINAL LAW 213 (Stephen Shute, John Gardner, & Jeremy Horder eds., 2003)

Alan Turing, *Computing Machinery and Intelligence*, 59 MIND 433 (1950)

AUSTIN TURK, CRIMINALITY AND LEGAL ORDER (1969)

HORSFALL J. TURNER, THE ANNALS OF THE WAKEFIELD HOUSE OF CORRECTIONS FOR THREE HUNDRED YEARS (1904)

ALAN TYREE, EXPERT SYSTEMS IN LAW (1989)

Mark S. Umbreit, *Community Service Sentencing: Jail Alternatives or Added Sanction?*, 45 FED. PROBATION 3 (1981)

Max L. Veech & Charles R. Moon, *De Minimis non Curat Lex*, 45 MICH. L. REV. 537 (1947)

RUSS VERSTEEG, EARLY MESOPOTAMIAN LAW (2000)

John Barker Waite, *The Law of Arrest*, 24 TEX. L. REV. 279 (1946)

NIGEL WALKER & NICOLA PADFIELD, SENTENCING: THEORY, LAW AND PRACTICE (1996)

Andrew Walkover, *The Infancy Defense in the New Juvenile Court*, 31 U.C.L.A. L. REV. 503 (1984)

Steven Walt & William S. Laufer, *Why Personhood Doesn't Matter: Corporate Criminal Liability and Sanctions*, 18 AM. J. CRIM. L. 263 (1991)

Mary Anne Warren, *On the Moral and Legal Status of Abortion*, in ETHICS IN PRACTICE (Hugh Lafollette ed., 1997)

DONALD A. WATERMAN, A GUIDE TO EXPERT SYSTEMS (1986)

MAX WEBER, ECONOMY AND SOCIETY: AN OUTLINE OF INTERPRETIVE SOCIOLOGY (1968)

HENRY WEIHOFEN, MENTAL DISORDER AS A CRIMINAL DEFENSE (1954)

Paul Weiss, *On the Impossibility of Artificial Intelligence*, 44 REV. METAPHYSICS 335 (1990)

Celia Wells, *Battered Woman Syndrome and Defences to Homicide: Where Now?*, 14 LEGAL STUD. 266 (1994)

Yueh-Hsuan Weng, Chien-Hsun Chen, & Chuen-Tsai Sun, *The Legal Crisis of Next Generation Robots: On Safety Intelligence*, Proceedings of the 11th International Conference on Artificial Intelligence and Law 205 (2007)

Yueh-Hsuan Weng, Chien-Hsun Chen, & Chuen-Tsai Sun, *Toward the Human-Robot Co-Existence Society: On Safety Intelligence for Next Generation Robots*, 1 INT. J. SOC. ROBOT. 267 (2009)

FRANCIS ANTONY WHITLOCK, CRIMINAL RESPONSIBILITY AND MENTAL ILLNESS (1963)

GLANVILLE WILLIAMS, CRIMINAL LAW: THE GENERAL PART (2d ed. 1961)

Glanville Williams, *Oblique Intention*, 46 CAMB. L.J. 417 (1987)

Glanville Williams, *The Draft Code and Reliance upon Official Statements*, 9 LEGAL STUD. 177 (1989)

Glanville Williams, *Innocent Agency and Causation*, 3 CRIM. L. F. 289 (1992)

Ashlee Willis, *Community Service as an Alternative to Imprisonment: A Cautionary View*, 24 PROBATION J. 120 (1977)

EDWARD O. WILSON, SOCIOBIOLOGY: THE NEW SYNTHESIS (1975)
JAMES Q. WILSON, THINKING ABOUT CRIME (2d ed. 1985)
TERRY WINOGRAD & FERNANDO C. FLORES, UNDERSTANDING COMPUTERS AND COGNITION: A NEW FOUNDATION FOR DESIGN (1986, 1987)
Terry Winograd, *Thinking Machines: Can There Be? Are We?*, in THE FOUNDATIONS OF ARTIFICIAL INTELLIGENCE 167 (Derek Pertridge & Yorick Wilks eds., 1990, 2006)
PATRICK HENRY WINSTON, ARTIFICIAL INTELLIGENCE (3d ed. 1992)
Edward M. Wise, *The Concept of Desert*, 33 WAYNE L. REV. 1343 (1987)
LUDWIG WITTGENSTEIN, PHILOSOPHISCHE UNTERSUCHUNGEN (1953)
Steven J. Wolhandler, *Voluntary Active Euthanasia for the Terminally Ill and the Constitutional Right to Privacy*, 69 CORNELL L. REV. 363 (1984)
Kam C. Wong, *Police Powers and Control in the People's Republic of China: The History of Shoushen*, 10 COLUM. J. ASIAN L. 367 (1996)
Ledger Wood, *Responsibility and Punishment*, 28 AM. INST. CRIM. L. & CRIMINOLOGY 630 (1938)
ANDREW WRIGHT, GWYNETH BOSWELL, & MARTIN DAVIES, CONTEMPORARY PROBATION PRACTICE (1993)
Andrew J. Wu, *From Video Games to Artificial Intelligence: Assigning Copyright Ownership to Works Generated by Increasingly Sophisticated Computer Programs*, 25 AIPLA Q. J. 131 (1997)
REUVEN YARON, THE LAWS OF ESHNUNNA (2d ed. 1988)
MASOUD YAZDANI & AJIT NARAYANAN, ARTIFICIAL INTELLIGENCE: HUMAN EFFECTS (1985)
PETER YOUNG, PUNISHMENT, MONEY AND THE LEGAL ORDER: AN ANALYSIS OF THE EMERGENCE OF MONETARY SANCTIONS WITH SPECIAL REFERENCE TO SCOTLAND (1987)
Rachel S. Zahniser, *Morally and Legally: A Parent's Duty to Prevent the Abuse of a Child as Defined by* Lane v. Commonwealth, 86 KY. L.J. 1209 (1998)
REINHARD ZIMMERMANN, THE LAW OF OBLIGATIONS — ROMAN FOUNDATIONS OF THE CIVILIAN TRADITION (1996)
Franklin E. Zimring, *The Executioner's Dissonant Song: On Capital Punishment and American Legal Values*, in THE KILLING STATE — CAPITAL PUNISHMENT IN LAW, POLITICS, AND CULTURE 137 (Austin Sarat ed., 1999)

Cases

Abrams v. United States, 250 U.S. 616, 63 L.Ed. 1173, 40 S.Ct. 17 (1919)
Adams v. State, 8 Md.App. 684, 262 A.2d 69 (1970)
Alford v. State, 866 S.W.2d 619 (Tex.Crim.App.1993)
Allday, (1837) 8 Car. & P. 136, 173 E.R. 431
Almon, (1770) 5 Burr. 2686, 98 E.R. 411
Anderson v. State, 66 Okl.Cr. 291, 91 P.2d 794 (1939)
Anderson, [1966] 2 Q.B. 110, [1966] 2 All E.R. 644, [1966] 2 W.L.R. 1195, 50 Cr. App. Rep. 216, 130 J.P. 318

SELECTED BIBLIOGRAPHY · 228

Andrews v. People, 800 P.2d 607 (Colo.1990)
Ann v. State, 30 Tenn. 159, 11 Hum. 159 (1850)
Arp v. State, 97 Ala. 5, 12 So. 301 (1893)
Axtell, (1660) 84 E.R. 1060
B. v. Director of Public Prosecutions, [2000] 2 A.C. 428, [2000] 1 All E.R. 833, [2000] 2 W.L.R. 452, [2000] 2 Cr. App. Rep. 65, [2000] Crim. L.R. 403
Bailey, (1818) Russ. & Ry. 341, 168 E.R. 835
Barnes v. State, 19 Conn. 398 (1849)
Barnfather v. Islington London Borough Council, [2003] E.W.H.C. 418 (Admin), [2003] 1 W.L.R. 2318, [2003] E.L.R. 263
Bateman, [1925] All E.R. Rep. 45, 94 L.J.K.B. 791, 133 L.T. 730, 89 J.P. 162, 41 T.L.R. 557, 69 Sol. Jo. 622, 28 Cox. C.C. 33, 19 Cr. App. Rep. 8
Batson v. State, 113 Nev. 669, 941 P.2d 478 (1997)
Beason v. State, 96 Miss. 165, 50 So. 488 (1909)
Benge, (1865) 4 F. & F. 504, 176 E.R. 665
Birmingham, &c., Railway Co., (1842) 3 Q. B. 223, 114 E.R. 492
Birney v. State, 8 Ohio Rep. 230 (1837)
Blake v. United States, 407 F.2d 908 (5th Cir.1969)
Blaker v. Tillstone, [1894] 1 Q.B. 345
Bolden v. State, 171 S.W.3d 785 (2005)
Bonder v. State, 752 A.2d 1169 (Del.2000)
Boson v. Sandford, (1690) 2 Salkeld 440, 91 E.R. 382
Boushea v. United States, 173 F.2d 131, 134 (8th Cir. 1949)
Bradley v. State, 102 Tex.Crim.R. 41, 277 S.W. 147 (1926)
Bratty v. Attorney-General for Northern Ireland, [1963] A.C. 386, 409, [1961] 3 All E.R. 523, [1961] 3 W.L.R. 965, 46 Cr. App. Rep 1
Brett v. Rigden, (1568) 1 Plowd. 340, 75 E.R. 516
Burnett, (1815) 4 M. & S. 272, 105 E.R. 835
C, [2007] E.W.C.A. Crim. 1862, [2007] All E.R. (D) 91
Caldwell, [1982] A.C. 341, [1981] 1 All E.R. 961, [1981] 2 W.L.R. 509, 73 Cr. App. Rep. 13, 145 J.P. 211
Calley v. Callaway, 519 F.2d 184 (5th Cir.1975)
Campbell v. Wood, 18 F.3d 662 (9th Cir. 1994)
Carter v. State, 376 P.2d 351 (Okl.Crim.App.1962)
Carter v. United States, 530 U.S. 255, 120 S.Ct. 2159, 147 L.Ed.2d 203 (2000)
Chance v. State, 685 A.2d 351 (Del.1996)
Cheek v. United States, 498 U.S. 192, 111 S.Ct. 604, 112 L.Ed.2d 617 (1991)
Childs v. State, 109 Nev. 1050, 864 P.2d 277 (1993)
Chisholm v. Doulton, (1889) 22 Q.B.D. 736
Chrystal v. Commonwealth, 72 Ky. 669, 9 Bush. 669 (1873)
City of Chicago v. Mayer, 56 Ill.2d 366, 308 N.E.2d 601 (1974)
Coal & C.R. v. Conley, 67 W.Va. 129, 67 S.E. 613 (1910)
Commonwealth v. Boynton, 84 Mass. 160, 2 Allen 160 (1861)
Commonwealth v. Fortner L.P. Gas Co., 610 S.W.2d 941 (Ky.App.1980)

Commonwealth v. French, 531 Pa. 42, 611 A.2d 175 (1992)
Commonwealth v. Goodman, 97 Mass. 117 (1867)
Commonwealth v. Green, 477 Pa. 170, 383 A.2d 877 (1978)
Commonwealth v. Herd, 413 Mass. 834, 604 N.E.2d 1294 (1992)
Commonwealth v. Hill, 11 Mass. 136 (1814)
Commonwealth v. Johnson, 412 Mass. 368, 589 N.E.2d 311 (1992)
Commonwealth v. Leno, 415 Mass. 835, 616 N.E.2d 453 (1993)
Commonwealth v. Lindsey, 396 Mass. 840, 489 N.E.2d 666 (1986)
Commonwealth v. McIlwain School Bus Lines, Inc., 283 Pa.Super. 1, 423 A.2d 413 (1980)
Commonwealth v. Mead, 92 Mass. 398 (1865)
Commonwealth v. Monico, 373 Mass. 298, 366 N.E.2d 1241 (1977)
Commonwealth v. New York Cent. & H. River R. Co., 206 Mass. 417, 92 N.E. 766 (1910)
Commonwealth v. Perl, 50 Mass.App.Ct. 445, 737 N.E.2d 937 (2000)
Commonwealth v. Pierce, 138 Mass. 165 (1884)
Commonwealth v. Proprietors of New Bedford Bridge, 68 Mass. 339 (1854)
Commonwealth v. Shumway, 72 Va.Cir. 481 (2007)
Commonwealth v. Thompson, 6 Mass. 134, 6 Tyng 134 (1809)
Commonwealth v. Walensky, 316 Mass. 383, 55 N.E.2d 902 (1944)
Commonwealth v. Weaver, 400 Mass. 612, 511 N.E.2d 545 (1987)
Cox v. State, 305 Ark. 244, 808 S.W.2d 306 (1991)
Crawshaw, (1860) Bell. 303, 169 E.R. 1271
Cutter v. State, 36 N.J.L. 125 (1873)
Da Silva, [2006] E.W.C.A. Crim. 1654 , [2006] 4 All E.R. 900, [2006] 2 Cr. App. Rep. 517
Dalloway, (1847) 2 Cox C.C. 273
Daniel v. State, 187 Ga. 411, 1 S.E.2d 6 (1939)
Director of Public Prosecutions v. Kent and Sussex Contractors Ltd., [1944] K.B. 146, [1944] 1 All E.R. 119
Dixon, (1814) 3 M. & S. 11, 105 E.R. 516
Dodd, (1736) Sess. Cas. 135, 93 E.R. 136
Dotson v. State, 6 Cold. 545 (1869)
Driver v. State, 2011 Tex. Crim. App. Lexis 4413 (2011)
Duckett v. State, 966 P.2d 941 (Wyo.1998)
Dudley and Stephens, [1884] 14 Q.B. D. 273
Dugdale, (1853) 1 El. & Bl. 435, 118 E.R. 499
Dusenbery v. Commonwealth, 263 S.E.2d 392 (Va. 1980)
Dutton v. State, 123 Md. 373, 91 A. 417 (1914)
Dyke v. Gower, [1892] 1 Q.B. 220
Elk v. United States, 177 U.S. 529, 20 S.Ct. 729, 44 L.Ed. 874 (1900)
English, [1999] A.C. 1, [1997] 4 All E.R. 545, [1997] 3 W.L.R. 959, [1998] 1 Cr. App. Rep. 261, [1998] Crim. L.R. 48, 162 J.P. 1
Esop, (1836) 7 Car. & P. 456, 173 E.R. 203

Evans v. Bartlam, [1937] A.C. 473, 479, [1937] 2 All E.R. 646
Evans v. State, 322 Md. 24, 585 A.2d 204 (1991)
Fain v. Commonwealth, 78 Ky. 183, 39 Am.Rep. 213 (1879)
Farmer v. People, 77 Ill. 322 (1875)
Finney, (1874) 12 Cox C.C. 625
Firth, (1990) 91 Cr. App. Rep. 217, 154 J.P. 576, [1990] Crim. L.R. 326
Fitzpatrick v. Kelly, [1873] 8 Q.B. 337
Fitzpatrick, [1977] N.I. 20
Forbes, (1835) 7 Car. & P. 224, 173 E.R. 99
Frey v. United States, 708 So.2d 918 (Fla.1998)
G., [2003] U.K.H.L. 50, [2004] 1 A.C. 1034, [2003] 3 W.L.R. 1060, [2003] 4 All E.R. 765, [2004] 1 Cr. App. Rep. 21, (2003) 167 J.P. 621, [2004] Crim. L. R. 369
G., [2008] U.K.H.L. 37, [2009] A.C. 92
Gammon (Hong Kong) Ltd. v. Attorney-General of Hong Kong, [1985] 1 A.C. 1, [1984] 2 All E.R. 503, [1984] 3 W.L.R. 437, 80 Cr. App. Rep. 194, 26 Build L.R. 159
Gardiner, [1994] Crim. L.R. 455
Godfrey v. State, 31 Ala. 323 (1858)
Government of the Virgin Islands v. Smith, 278 F.2d 169 (3rd Cir.1960)
Granite Construction Co. v. Superior Court, 149 Cal.App.3d 465, 197 Cal. Rptr. 3 (1983)
Gray v. Lucas, 710 F.2d 1048 (5th Cir. 1983)
Great Broughton (Inhabitants), (1771) 5 Burr. 2700, 98 E.R. 418
Gregg v. Georgia, 428 U.S. 153, S.Ct. 2909, 49 L.Ed.2d 859 (1979)
Grout, (1834) 6 Car. & P. 629, 172 E.R. 1394
Hall v. Brooklands Auto Racing Club, [1932] All E.R. 208, [1933] 1 K.B. 205, 101 L.J.K.B. 679, 147 L.T. 404, 48 T.L.R. 546
Hardcastle v. Bielby, [1892] 1 Q.B. 709
Hartson v. People, 125 Colo. 1, 240 P.2d 907 (1951)
Hasan, [2005] U.K.H.L. 22, [2005] 4 All E.R. 685, [2005] 2 Cr. App. Rep. 314, [2006] Crim. L.R. 142, [2005] All E.R. (D) 299
Heilman v. Commonwealth, 84 Ky. 457, 1 S.W. 731 (1886)
Henderson v. Kibbe, 431 U.S. 145, 97 S.Ct. 1730, 52 L.Ed.2d 203 (1977)
Henderson v. State, 11 Ala.App. 37, 65 So. 721 (1914)
Hentzner v. State, 613 P.2d 821 (Alaska 1980)
Hern v. Nichols, (1708) 1 Salkeld 289, 91 E.R. 256
Hobbs v. Winchester Corporation, [1910] 2 K.B. 471
Holbrook, (1878) 4 Q.B.D. 42
Howard v. State, 73 Ga.App. 265, 36 S.E.2d 161 (1945)
Huggins, (1730) 2 Strange 882, 93 E.R. 915
Hughes v. Commonwealth, 19 Ky.L.R. 497, 41 S.W. 294 (1897)
Hull, (1664) Kel. 40, 84 E.R. 1072
Humphrey v. Commonwealth, 37 Va.App. 36, 553 S.E.2d 546 (2001)

SELECTED BIBLIOGRAPHY · 231

Hunt v. Nuth, 57 F.3d 1327 (4th Cir. 1995)
Hunt v. State, 753 So.2d 609 (Fla.App.2000)
Hunter v. State, 30 Tenn. 160, 1 Head 160 (1858)
I.C.R. Haulage Ltd., [1944] K.B. 551, [1944] 1 All E.R. 691
Ingram v. United States, 592 A.2d 992 (D.C.App.1991)
Johnson v. State, 142 Ala. 70 (1904)
Jones v. Hart, (1699) 2 Salkeld 441, 91 E.R. 382
Jurco v. State, 825 P.2d 909 (Alaska App.1992)
K., [2001] U.K.H.L. 41, [2002] 1 A.C. 462
Kimoktoak v. State, 584 P.2d 25 (Alaska 1978)
Kingston v. Booth, (1685) Skinner 228, 90 E.R. 105
Knight, (1828) 1 L.C.C. 168, 168 E.R. 1000
Kumar, [2004] E.W.C.A. Crim. 3207, [2005] 1 Cr. App. Rep. 566, [2005] Crim. L.R. 470
Laaman v. Helgemoe, 437 F.Supp. 269 (1977)
Lambert v. California, 355 U.S. 225, 78 S.Ct. 240, 2 L.Ed.2d 228 (1957)
Lambert v. State, 374 P.2d 783 (Okla.Crim.App.1962)
Lane v. Commonwealth, 956 S.W.2d 874 (Ky.1997)
Langforth Bridge, (1635) Cro. Car. 365, 79 E.R. 919
Larsonneur, (1933) 24 Cr. App. R. 74, 97 J.P. 206, 149 L.T. 542
Lawson, [1986] V.R. 515
Leach, [1937] 1 All E.R. 319
Lee v. State, 41 Tenn. 62, 1 Cold. 62 (1860)
Leet v. State, 595 So.2d 959 (1991)
Lennard's Carrying Co. Ltd. v. Asiatic Petroleum Co. Ltd., [1915] A.C. 705
In re Leroy, 285 Md. 508, 403 A.2d 1226 (1979)
Lester v. State, 212 Tenn. 338, 370 S.W.2d 405 (1963)
Levett, (1638) Cro. Car. 538
lifton (Inhabitants), (1794) 5 T.R. 498, 101 E.R. 280
Liverpool (Mayor), (1802) 3 East 82, 102 E.R. 529
Long v. Commonwealth, 23 Va.App. 537, 478 S.E.2d 324 (1996)
Long v. State, 44 Del. 262, 65 A.2d 489 (1949)
Longbottom, (1849) 3 Cox C. C. 439
Lutwin v. State, 97 N.J.L. 67, 117 A. 164 (1922)
Manser, (1584) 2 Co. Rep. 3, 76 E.R. 392
Marshall, (1830) 1 Lewin 76, 168 E.R. 965
Martin v. State, 90 Ala. 602, 8 So. 858 (1891)
Mason v. State, 603 P.2d 1146 (Okl.Crim.App.1979)
Matudi, [2004] E.W.C.A. Crim. 697
Mavji, [1987] 2 All E.R. 758, [1987] 1 W.L.R. 1388, [1986] S.T.C. 508, Cr. App. Rep. 31, [1987] Crim. L.R. 39
Maxey v. United States, 30 App. D.C. 63, 80 (App. D.C. 1907)
McClain v. State, 678 N.E.2d 104 (Ind.1997)
McGrowther, (1746) 18 How. St. Tr. 394

McMillan v. City of Jackson, 701 So.2d 1105 (Miss.1997)
McNeil v. United States, 933 A.2d 354 (2007)
Meade, [1909] 1 K.B. 895
Meakin, (1836) 7 Car. & P. 297, 173 E.R. 131
Mendez v. State, 575 S.W.2d 36 (Tex.Crim.App.1979)
Michael, (1840) 2 Mood. 120, 169 E.R. 48
Middleton v. Fowler, (1699) 1 Salkeld 282, 91 E.R. 247
Mildmay, (1584) 1 Co. Rep. 175a, 76 E.R. 379
Miller v. State, 3 Ohio St. Rep. 475 (1854)
Minor v. State, 326 Md. 436, 605 A.2d 138 (1992)
Mitchell v. State, 114 Nev. 1417, 971 P.2d 813 (1998)
M'Naghten, (1843) 10 Cl. & Fin. 200, 8 E.R. 718
Montgomery v. Commonwealth, 189 Ky. 306, 224 S.W. 878 (1920)
Moore v. State, 25 Okl.Crim. 118, 218 P. 1102 (1923)
Mouse, (1608) 12 Co. Rep. 63, 77 E.R. 1341
Myers v. State, 1 Conn. 502 (1816)
Nelson v. State, 597 P.2d 977 (Alaska 1979)
New York & G.L.R. Co. v. State, 50 N.J.L. 303, 13 A. 1 (1888)
New York Cent. & H.R.R. v. United States, 212 U.S. 481, 29 S.Ct. 304, 53 L.Ed. 613 (1909)
Nutt, (1728) 1 Barn. K.B. 306, 94 E.R. 208
O'Flaherty, [2004] E.W.C.A. Crim. 526, [2004] 2 Cr. App. Rep. 315
Oxford, (1840) 9 Car. & P. 525, 173 E.R. 941
Parish, (1837) 8 Car. & P. 94, 173 E.R. 413
Parnell v. State, 912 S.W.2d 422, 424 (Ark. 1996)
Pearson, (1835) 2 Lewin 144, 168 E.R. 1108
Peebles v. State, 101 Ga. 585, 28 S.E. 920 (1897)
People v. Bailey, 451 Mich. 657, 549 N.W.2d 325 (1996)
People v. Brubaker, 53 Cal.2d 37, 346 P.2d 8 (1959)
People v. Cabaltero, 31 Cal.App.2d 52, 87 P.2d 364 (1939)
People v. Cherry, 307 N.Y. 308, 121 N.E.2d 238 (1954)
People v. Clark, 8 N.Y.Cr. 169, 14 N.Y.S. 642 (1891)
People v. Cooper, 194 Ill.2d 419, 252 Ill.Dec. 458, 743 N.E.2d 32 (2000)
People v. Craig, 78 N.Y.2d 616, 578 N.Y.S.2d 471, 585 N.E.2d 783 (1991)
People v. Daugherty, 40 Cal.2d 876, 256 P.2d 911 (1953)
People v. Davis, 33 N.Y.2d 221, 351 N.Y.S.2d 663, 306 N.E.2d 787 (1973)
People v. Decina, 2 N.Y.2d 133, 157 N.Y.S.2d 558, 138 N.E.2d 799 (1956)
People v. Disimone, 251 Mich.App. 605, 650 N.W.2d 436 (2002)
People v. Ferguson, 134 Cal.App. 41, 24 P.2d 965 (1933)
People v. Freeman, 61 Cal.App.2d 110, 142 P.2d 435 (1943)
People v. Handy, 198 Colo. 556, 603 P.2d 941 (1979)
People v. Haney, 30 N.Y.2d 328, 333 N.Y.S.2d 403, 284 N.E.2d 564 (1972)
People v. Harris, 29 Cal. 678 (1866)
People v. Heitzman, 9 Cal.4th 189, 37 Cal.Rptr.2d 236, 886 P.2d 1229 (1994)

People v. Henry, 239 Mich.App. 140, 607 N.W.2d 767 (1999)
People v. Higgins, 5 N.Y.2d 607, 186 N.Y.S.2d 623, 159 N.E.2d 179 (1959)
People v. Howk, 56 Cal.2d 687, 16 Cal.Rptr. 370, 365 P.2d 426 (1961)
People v. Kemp, 150 Cal.App.2d 654, 310 P.2d 680 (1957)
People v. Kessler, 57 Ill.2d 493, 315 N.E.2d 29 (1974)
People v. Kirst, 168 N.Y. 19, 60 N.E. 1057 (1901)
People v. Larkins, 2010 Mich. App. Lexis 1891 (2010)
People v. Leonardi, 143 N.Y. 360, 38 N.E. 372 (1894)
People v. Lisnow, 88 Cal.App.3d Supp. 21, 151 Cal.Rptr. 621 (1978)
People v. Little, 41 Cal.App.2d 797, 107 P.2d 634 (1941)
People v. Marshall, 362 Mich. 170, 106 N.W.2d 842 (1961)
People v. Merhige, 212 Mich. 601, 180 N.W. 418 (1920)
People v. Michalow, 229 N.Y. 325, 128 N.E. 228 (1920)
People v. Minifie, 13 Cal.4th 1055, 56 Cal.Rptr.2d 133, 920 P.2d 1337 (1996)
People v. Monks, 133 Cal. App. 440 (Cal. Dist. Ct. App. 1933)
People v. Mutchler, 140 N.E. 820, 823 (Ill. 1923)
People v. Newton, 8 Cal.App.3d 359, 87 Cal.Rptr. 394 (1970)
People v. Pantano, 239 N.Y. 416, 146 N.E. 646 (1925)
People v. Prettyman, 14 Cal.4th 248, 58 Cal.Rptr.2d 827, 926 P.2d 1013 (1996)
People v. Richards, 269 Cal.App.2d 768, 75 Cal.Rptr. 597 (1969)
People v. Sakow, 45 N.Y.2d 131, 408 N.Y.S.2d 27, 379 N.E.2d 1157 (1978)
People v. Smith, 57 Cal. App. 4th 1470, 67 Cal. Rptr. 2d 604 (1997)
People v. Sommers, 200 P.3d 1089 (2008)
People v. Townsend, 214 Mich. 267, 183 N.W. 177 (1921)
People v. Vogel, 46 Cal.2d 798, 299 P.2d 850 (1956)
People v. Weiss, 256 App.Div. 162, 9 N.Y.S.2d 1 (1939)
People v. Whipple, 100 Cal.App. 261, 279 P. 1008 (1929)
People v. Williams, 56 Ill.App.2d 159, 205 N.E.2d 749 (1965)
People v. Wilson, 66 Cal.2d 749, 59 Cal.Rptr. 156, 427 P.2d 820 (1967)
People v. Young, 11 N.Y.2d 274, 229 N.Y.S.2d 1, 183 N.E.2d 319 (1962)
Pierson v. State, 956 P.2d 1119 (Wyo.1998)
Pigman v. State, 14 Ohio 555 (1846)
Polston v. State, 685 P.2d 1 (Wyo.1984)
Pope v. United States, 372 F.2d 710 (8th Cir.1970)
Powell v. Texas, 392 U.S. 514, 88 S.Ct. 2145, 20 L.Ed.2d 1254 (1968)
Price v. State, 50 Tex.Crim.R. 71, 94 S.W. 901 (1906)
Proctor v. State, 15 Okl.Cr. 338, 176 P. 771 (1918)
Provenzano v. Moore, 744 So.2d 413 (Fla. 1999)
Provincial Motor Cab Company Ltd. v. Dunning, [1909] 2 K.B. 599
Pugliese v. Commonwealth, 16 Va.App. 82, 428 S.E.2d 16 (1993)
Quick, [1973] Q.B. 910, [1973] 3 All E.R. 347, [1973] 3 W.L.R. 26, 57 Cr. App. Rep. 722, 137 J.P. 763
R.I. Recreation Center v. Aetna Cas. & Surety Co., 177 F.2d 603 (1st Cir.1949)
Rangel v. State, 2009 Tex.App. 1555 (2009)

Ratzlaf v. United States, 510 U.S. 135, 114 S.Ct. 655, 126 L.Ed.2d 615 (1994)
Read v. People, 119 Colo. 506, 205 P.2d 233 (1949)
Redmond v. State, 36 Ark. 58 (1880)
Reed v. State, 693 N.E.2d 988 (Ind.App.1998)
Reniger v. Fogossa, (1551) 1 Plowd. 1, 75 E.R. 1
Rice v. State, 8 Mo. 403 (1844)
Richards, [2004] E.W.C.A. Crim. 192
Richardson v. State, 697 N.E.2d 462 (Ind.1998)
Ricketts v. State, 291 Md. 701, 436 A.2d 906 (1981)
Roberts v. People, 19 Mich. 401 (1870)
Robinson v. California, 370 U.S. 660, 82 S.Ct. 1417, 8 L.Ed.2d 758 (1962)
Rollins v. State, 2009 Ark. 484, 347 S.W.3d 20 (2009)
Roy v. United States, 652 A.2d 1098 (D.C.App.1995)
Saik, [2006] U.K.H.L. 18, [2007] 1 A.C. 18
Saintiff, (1705) 6 Mod. 255, 87 E.R. 1002
Sam v. Commonwealth, 13 Va.App. 312, 411 S.E.2d 832 (1991)
Sanders v. State, 466 N.E.2d 424 (Ind.1984)
Scales v. United States, 367 U.S. 203, 81 S.Ct. 1469, 6 L.Ed.2d 782 (1961)
Schmidt v. United States, 133 F. 257 (9th Cir.1904)
Schuster v. State, 48 Ala. 199 (1872)
Scott v. State, 71 Tex.Crim.R. 41, 158 S.W. 814 (1913)
Seaboard Offshore Ltd. v. Secretary of State for Transport, [1994] 2 All E.R. 99, [1994] 1 W.L.R. 541, [1994] 1 Lloyd's Rep. 593
Seaman v. Browning, (1589) 4 Leonard 123, 74 E.R. 771
Severn and Wye Railway Co., (1819) 2 B. & Ald. 646, 106 E.R. 501
Ex parte Smith, 135 Mo. 223, 36 S.W. 628 (1896)
Smith v. California, 361 U.S. 147, 80 S.Ct. 215, 4 L.Ed.2d 205 (1959)
Smith v. State, 83 Ala. 26, 3 So. 551 (1888)
Spiers & Pond v. Bennett, [1896] 2 Q.B. 65
Squire v. State, 46 Ind. 459 (1874)
State Philbrick, 402 A.2d 59 (Me.1979)
State v. Aaron, 4 N.J.L. 269 (1818)
State v. Anderson, 141 Wash.2d 357, 5 P.3d 1247 (2000)
State v. Anthuber, 201 Wis.2d 512, 549 N.W.2d 477 (App.1996)
State v. Asher, 50 Ark. 427, 8 S.W. 177 (1888)
State v. Audette, 149 Vt. 218, 543 A.2d 1315 (1988)
State v. Ayer, 136 N.H. 191, 612 A.2d 923 (1992)
State v. Barker, 128 W.Va. 744, 38 S.E.2d 346 (1946)
State v. Barrett, 768 A.2d 929 (R.I.2001)
State v. Blakely, 399 N.W.2d 317 (S.D.1987)
State v. Bono, 128 N.J.Super. 254, 319 A.2d 762 (1974)
State v. Bowen, 118 Kan. 31, 234 P. 46 (1925)
State v. Brosnan, 221 Conn. 788, 608 A.2d 49 (1992)
State v. Brown, 389 So.2d 48 (La.1980)

State v. Bunkley, 202 Conn. 629, 522 A.2d 795 (1987)
State v. Burrell, 135 N.H. 715, 609 A.2d 751 (1992)
State v. Cain, 9 W. Va. 559 (1874)
State v. Cameron, 104 N.J. 42, 514 A.2d 1302 (1986)
State v. Campbell, 536 P.2d 105 (Alaska 1975)
State v. Carrasco, 122 N.M. 554, 928 P.2d 939 (1996)
State v. Case, 672 A.2d 586 (Me.1996)
State v. Caswell, 771 A.2d 375 (Me.2001)
State v. Champa, 494 A.2d 102 (R.I.1985)
State v. Chicago, M. & St.P.R. Co., 130 Minn. 144, 153 N.W. 320 (1915)
State v. Clottu, 33 Ind. 409 (1870)
State v. Coffin, 128 N.M. 192, 991 P.2d 477 (1999)
State v. Cram, 157 Vt. 466, 600 A.2d 733 (1991)
State v. Crocker, 431 A.2d 1323 (Me.1981)
State v. Crocker, 506 A.2d 209 (Me.1986)
State v. Cude, 14 Utah 2d 287, 383 P.2d 399 (1963)
State v. Curry, 45 Ohio St.3d 109, 543 N.E.2d 1228 (1989)
State v. Daniels, 236 La. 998, 109 So.2d 896 (1958)
State v. Dansinger, 521 A.2d 685 (Me.1987)
State v. Daoud, 141 N.H. 142, 679 A.2d 577 (1996)
State v. Dillon, 93 Idaho 698, 471 P.2d 553 (1970)
State v. Dubina, 164 Conn. 95, 318 A.2d 95 (1972)
State v. Ehlers, 98 N.J.L. 263, 119 A. 15 (1922)
State v. Ellis, 232 Or. 70, 374 P.2d 461 (1962)
State v. Elsea, 251 S.W.2d 650 (Mo.1952)
State v. Etzweiler, 125 N.H. 57, 480 A.2d 870 (1984)
State v. Evans, 134 N.H. 378, 594 A.2d 154 (1991)
State v. Farley, 225 Kan. 127, 587 P.2d 337 (1978)
State v. Fee, 126 N.H. 78, 489 A.2d 606 (1985)
State v. Finnell, 101 N.M. 732, 688 P.2d 769 (1984)
State v. Fletcher, 322 N.C. 415, 368 S.E.2d 633 (1988)
State v. Follin, 263 Kan. 28, 947 P.2d 8 (1997)
State v. Foster, 202 Conn. 520, 522 A.2d 277 (1987)
State v. Foster, 91 Wash.2d 466, 589 P.2d 789 (1979)
State v. Gallagher, 191 Conn. 433, 465 A.2d 323 (1983)
State v. Gartland, 304 Mo. 87, 263 S.W. 165 (1924)
State v. Garza, 259 Kan. 826, 916 P.2d 9 (1996)
State v. George, 20 Del. 57, 54 A. 745 (1902)
State v. Gish, 17 Idaho 341, 393 P.2d 342 (1964)
State v. Goodall, 407 A.2d 268 (Me.1979)
State v. Goodenow, 65 Me. 30 (1876)
State v. Gray, 221 Conn. 713, 607 A.2d 391 (1992)
State v. Great Works Mill. & Mfg. Co., 20 Me. 41, 37 Am.Dec.38 (1841)
State v. Hadley, 65 Utah 109, 234 P. 940 (1925)

SELECTED BIBLIOGRAPHY · 236

State v. Harris, 222 N.W.2d 462 (Iowa 1974)
State v. Hastings, 118 Idaho 854, 801 P.2d 563 (1990)
State v. Havican, 213 Conn. 593, 569 A.2d 1089 (1990)
State v. Herro, 120 Ariz. 604, 587 P.2d 1181 (1978)
State v. Hinkle, 200 W.Va. 280, 489 S.E.2d 257 (1996)
State v. Hobbs, 252 Iowa 432, 107 N.W.2d 238 (1961)
State v. Hooker, 17 Vt. 658 (1845)
State v. Hopkins, 147 Wash. 198, 265 P. 481 (1928)
State v. Howley, 128 Idaho 874, 920 P.2d 391 (1996)
State v. I. & M. Amusements, Inc., 10 Ohio App.2d 153, 226 N.E.2d 567 (1966)
State v. J.P.S., 135 Wash.2d 34, 954 P.2d 894 (1998)
State v. Jackson, 137 Wash.2d 712, 976 P.2d 1229 (1999)
State v. Jackson, 346 Mo. 474, 142 S.W.2d 45 (1940)
State v. Jacobs, 371 So.2d 801 (La.1979)
State v. Jenner, 451 N.W.2d 710 (S.D.1990)
State v. Johnson, 233 Wis. 668, 290 N.W. 159 (1940)
State v. Kaiser, 260 Kan. 235, 918 P.2d 629 (1996)
State v. Kee, 398 A.2d 384 (Me.1979)
State v. Labato, 7 N.J. 137, 80 A.2d 617 (1951)
State v. Lawrence, 97 N.C. 492, 2 S.E. 367 (1887)
State v. Linscott, 520 A.2d 1067 (Me.1987)
State v. Lockhart, 208 W.Va. 622, 542 S.E.2d 443 (2000)
State v. Lucas, 55 Iowa 321, 7 N.W. 583 (1880)
State v. Marley, 54 Haw. 450, 509 P.2d 1095 (1973)
State v. Martin, 119 N.J. 2, 573 A.2d 1359 (1990)
State v. McDowell, 312 N.W.2d 301 (N.D. 1981)
State v. Mendoza, 709 A.2d 1030 (R.I.1998)
State v. Mishne, 427 A.2d 450 (Me.1981)
State v. Molin, 288 N.W.2d 232 (Minn.1979)
State v. Moore, 158 N.J. 292, 729 A.2d 1021 (1999)
State v. Murphy, 674 P.2d 1220 (Utah.1983)
State v. Nargashian, 26 R.I. 299, 58 A. 953 (1904)
State v. Nelson, 329 N.W.2d 643 (Iowa 1983)
State v. Neuzil, 589 N.W.2d 708 (Iowa 1999)
State v. Nickelson, 45 La.Ann. 1172, 14 So. 134 (1893)
State v. Pereira, 72 Conn. App. 545, 805 A.2d 787 (2002)
State v. Pincus, 41 N.J.Super. 454, 125 A.2d 420 (1956)
State v. Reed, 205 Neb. 45, 286 N.W.2d 111 (1979)
State v. Reese, 272 N.W.2d 863 (Iowa 1978)
State v. Robinson, 132 Ohio App.3d 830, 726 N.E.2d 581 (1999)
State v. Robinson, 20 W.Va. 713, 43 Am.Rep. 799 (1882)
State v. Rocheville, 310 S.C. 20, 425 S.E.2d 32 (1993)
State v. Rocker, 52 Haw. 336, 475 P.2d 684 (1970)
State v. Runkles, 605 A.2d 111, 121 (Md. 1992)

State v. Sargent, 156 Vt. 463, 594 A.2d 401 (1991)
State v. Sasse, 6 S.D. 212, 60 N.W. 853 (1894)
State v. Sawyer, 95 Conn. 34, 110 A. 461 (1920)
State v. Schulz, 55 Ia. 628 (1881)
State v. Sexton, 160 N.J. 93, 733 A.2d 1125 (1999)
State v. Sheedy, 125 N.H. 108, 480 A.2d 887 (1984)
State v. Silva-Baltazar, 125 Wash.2d 472, 886 P.2d 138 (1994)
State v. Silveira, 198 Conn. 454, 503 A.2d 599 (1986)
State v. Smith, 170 Wis.2d 701, 490 N.W.2d 40 (App.1992)
State v. Smith, 219 N.W.2d 655 (Iowa 1974)
State v. Smith, 260 Or. 349, 490 P.2d 1262 (1971)
State v. Stepniewski, 105 Wis.2d 261, 314 N.W.2d 98 (1982)
State v. Stewart, 624 N.W.2d 585 (Minn.2001)
State v. Stoehr, 134 Wis.2d 66, 396 N.W.2d 177 (1986)
State v. Striggles, 202 Iowa 1318, 210 N.W. 137 (1926)
State v. Strong, 294 N.W.2d 319 (Minn.1980)
State v. Thomas, 619 S.W.2d 513, 514 (Tenn. 1981)
State v. Torphy, 78 Mo.App. 206 (1899)
State v. Torres, 495 N.W.2d 678 (1993)
State v. Totman, 80 Mo.App. 125 (1899)
State v. VanTreese, 198 Iowa 984, 200 N.W. 570 (1924)
State v. Warshow, 138 Vt. 22, 410 A.2d 1000 (1979)
State v. Welsh, 8 Wash.App. 719, 508 P.2d 1041 (1973)
State v. Wenger, 58 Ohio St.2d 336, 390 N.E.2d 801 (1979)
State v. Whitman, 116 Fla. 196, 156 So. 705 (1934)
State v. Whitoomb, 52 Iowa 85, 2 N.W. 970 (1879)
State v. Wilchinski, 242 Conn. 211, 700 A.2d 1 (1997)
State v. Wilson, 267 Kan. 550, 987 P.2d 1060 (1999)
State v. Wyatt, 198 W.Va. 530, 482 S.E.2d 147 (1996)
Stein v. State, 37 Ala. 123 (1861)
Stephens, [1866] 1 Q.B. 702
Stratford-upon-Avon Corporation, (1811) 14 East 348, 104 E.R. 636
Studstill v. State, 7 Ga. 2 (1849)
Sweet v. Parsley, [1970] A.C. 132, [1969] 1 All E.R. 347, [1969] 2 W.L.R. 470, 133 J.P. 188, 53 Cr. App. Rep. 221, 209 E.G. 703, [1969] E.G.D. 123
Tate v. Commonwealth, 258 Ky. 685, 80 S.W.2d 817 (1935)
Taylor v. State, 158 Miss. 505, 130 So. 502 (1930)
Texaco Inc. v. Short, 454 U.S. 516, 102 S.Ct. 781, 70 L.Ed.2d 738 (1982)
Thomas, (1837) 7 Car. & P. 817, 173 E.R. 356
Thompson v. State, 44 S.W.3d 171 (Tex.App.2001)
Thompson v. United States, 348 F.Supp.2d 398 (2005)
Tift v. State, 17 Ga.App. 663, 88 S.E. 41 (1916)
Treacy v. Director of Public Prosecutions, [1971] A.C. 537, 559, [1971] 1 All E.R. 110, [1971] 2 W.L.R. 112, 55 Cr. App. Rep. 113, 135 J.P. 112

Tully v. State, 730 P.2d 1206 (Okl.Crim.App.1986)
Turberwill v. Stamp, (1697) Skinner 681, 90 E.R. 303
In re Tyvonne, 211 Conn. 151, 558 A.2d 661 (1989)
United States v. Alaska Packers' Association, 1 Alaska 217 (1901)
United States v. Albertini, 830 F.2d 985 (9th Cir.1987)
United States v. Allegheny Bottling Company, 695 F.Supp. 856 (1988)
United States v. Andrews, 75 F.3d 552 (9th Cir.1996)
United States v. Arthurs, 73 F.3d 444 (1st Cir.1996)
United States v. Bailey, 444 U.S. 394, 100 S.Ct. 624, 62 L.Ed.2d 575 (1980)
United States v. Bakhtiari, 913 F.2d 1053 (2nd Cir.1990)
United States v. Bryan, 483 F.2d 88, 92 (3d Cir. 1973)
United States v. Buber, 62 M.J. 476 (2006)
United States v. Calley, 48 C.M.R. 19, 22 U.S.C.M.A. 534 (1973)
United States v. Campbell, 675 F.2d 815 (6th Cir.1982)
United States v. Carter, 311 F.2d 934 (6th Cir.1963)
United States v. Chandler, 393 F.2d 920 (4th Cir.1968)
United States v. Contento-Pachon, 723 F.2d 691 (9th Cir.1984)
United States v. Currens, 290 F.2d 751 (3rd Cir.1961)
United States v. Doe, 136 F.3d 631 (9th Cir.1998)
United States v. Dominguez-Ochoa, 386 F.3d 639 (2004)
United States v. Dorrell, 758 F.2d 427 (9th Cir.1985)
United States v. Dye Construction Co., 510 F.2d 78 (10th Cir.1975)
United States v. Freeman, 25 Fed. Cas. 1208 (1827)
United States v. Freeman, 357 F.2d 606 (2nd Cir.1966)
United States v. Gomez, 81 F.3d 846 (9th Cir.1996)
United States v. Greer, 467 F.2d 1064 (7th Cir.1972)
United States v. Hanousek, 176 F.3d 1116 (9th Cir.1999)
United States v. Heredia, 483 F.3d 913 (2006)
United States v. Holmes, 26 F. Cas. 360, 1 Wall. Jr. 1 (1842)
United States v. Jewell, 532 F.2d 697 (9th Cir.1976)
United States v. John Kelso Co., 86 F. 304 (Cal.1898)
United States v. Johnson, 956 F.2d 894 (9th Cir.1992)
United States v. Kabat, 797 F.2d 580 (8th Cir.1986)
United States v. Ladish Malting Co., 135 F.3d 484 (7th Cir.1998)
United States v. LaFleur, 971 F.2d 200 (9th Cir.1991)
United States v. Lampkins, 4 U.S.C.M.A. 31, 15 C.M.R. 31 (1954)
United States v. Lee, 694 F.2d 649 (11th Cir.1983)
United States v. Mancuso, 139 F.2d 90 (3rd Cir.1943)
United States v. Maxwell, 254 F.3d 21 (1st Cir.2001)
United States v. Meyers, 906 F. Supp. 1494 (1995)
United States v. Moore, 486 F.2d 1139 (D.C.Cir.1973)
United States v. Oakland Cannabis Buyers' Cooperative, 532 U.S. 483, 121 S.Ct. 1711, 149 L.Ed.2d 722 (2001)
United States v. Paolello, 951 F.2d 537 (3rd Cir.1991)

United States v. Pomponio, 429 U.S. 10, 97 S.Ct. 22, 50 L.Ed.2d 12 (1976)
United States v. Powell, 929 F.2d 724 (D.C.Cir.1991)
United States v. Quaintance, 471 F. Supp.2d 1153 (2006)
United States v. Ramon-Rodriguez, 492 F.3d 930 (2007)
United States v. Randall, 104 Wash.D.C.Rep. 2249 (D.C.Super.1976)
United States v. Randolph, 93 F.3d 656 (9th Cir.1996)
United States v. Robertson, 33 M.J. 832 (1991)
United States v. Ruffin, 613 F.2d 408 (2d Cir. 1979)
United States v. Shapiro, 383 F.2d 680 (7th Cir.1967)
United States v. Smith, 404 F.2d 720 (6th Cir.1968)
United States v. Spinney, 65 F.3d 231 (1st Cir.1995)
United States v. Sued-Jimenez, 275 F.3d 1 (1st Cir.2001)
United States v. Thompson-Powell Drilling Co., 196 F.Supp. 571 (N.D.Tex.1961)
United States v. Tobon-Builes, 706 F.2d 1092 (11th Cir. 1983)
United States v. Torres, 977 F.2d 321 (7th Cir.1992)
United States v. Warner, 28 Fed. Cas. 404, 6 W.L.J. 255, 4 McLean 463 (1848)
United States v. Wert-Ruiz, 228 F.3d 250 (3rd Cir.2000)
United States v. Youts, 229 F.3d 1312 (10th Cir.2000)
Vantandillo, (1815) 4 M. & S. 73, 105 E.R. 762
Vaux, (1613) 1 Blustrode 197, 80 E.R. 885
Virgin Islands v. Joyce, 210 F. App. 208 (2006)
Walter, (1799) 3 Esp. 21, 170 E.R. 524
Webb, [2006] E.W.C.A. Crim. 2496, [2007] All E.R. (D) 406
In re Welfare of C.R.M., 611 N.W.2d 802 (Minn.2000)
Wheatley v. Commonwealth, 26 Ky.L.Rep. 436, 81 S.W. 687 (1904)
Wieland v. State, 101 Md. App. 1, 643 A.2d 446 (1994)
Wilkerson v. Utah, 99 U.S. (9 Otto) 130, 25 L.Ed. 345 (1878)
Willet v. Commonwealth, 76 Ky. 230 (1877)
Williams v. State, 70 Ga.App. 10, 27 S.E.2d 109 (1943)
Williamson, (1807) 3 Car. & P. 635, 172 E.R. 579
Wilson v. State, 24 S.W. 409 (Tex.Crim.App.1893)
Wilson v. State, 777 S.W.2d 823 (Tex.App.1989)
In re Winship, 397 U.S. 358, 90 S.Ct. 1068, 25 L.Ed.2d 368 (1970)
Woodrow, (1846) 15 M. & W. 404, 153 E.R. 907

INDEX

Note: Page numbers in *italics* refer to tables

absolute defense model, 150
absolute responsibility model, 150
accessoryship, 29, 41, 69
accountability, 17–19
act, 39–41
actio libera in causa, 127, 141, 145, 148
animals, 7, 22–25
Asimov, 15–17, 141, 143
attention, 2, 50, 54
awareness: and AI systems, 125; cognition and volition, 31–33, *33*, 131–136, 154; definition of, in criminal law, 49–56, 63–64, 66–67, 98–100; and *mens rea*, 45–48, *49*, 98, 120–121; and negligence, 86–89, *89*, 90–91, 94–95, 98, 101–102, 118; self-awareness, 8, 127–128; and will, 58, 71–72, 82. See also *mens rea;* negligence

Babbage, 2
blindness, 52, 55–56. *See also* willful blindness
bona fide, 137, 139
bureaucracy, 10

capital punishment, 142, 157, 162, 165–167, 169
characteristica universalis, 1
chess, 2, 4, 11, 59–60
chinese room, 6–7

circumstances: and duress, 147; and infancy, 126; and necessity, 147; and negligence, 85, 88–89, *89*, 91–92, 102; and mental element requirement, 45–49, *49*, 52, 55–56; optional component, 38; and perpetration-through-another, 72; and probable consequence liability, 79–80; and self-defense, 141, 144; and sentencing, 156, 163, 167, 175; and strict liability, 105, 108, *109*; structure of factual element requirement, 30–32, *31*, 42–43; and substantive immunity, 139
cognition, 31–33, *33*, 45–48, *49*, 51, 54, 61, 88, 136
combat decisions, 4
combinatorial explosion, 10
commonsense, 3
communication, 7, 9, 24
community service, 171
complicity, 29, 68–69, 71, 78, 99
conduct: and AI entity liability, 65; and general defenses, 121, 129, 147; goal driven, 8–9; and mental element requirement, 38–49, *49*, 55–60, 63; and negligence, 85–89, *89*, 92–93; and perpetration-through-another, 69, 72–73, 114; and probable consequence liability, 76; and sentencing, 169; and strict liability, 105–108, *109*, 118; structure of factual element requirement, 26–32, *31*

corporations: and general criminal liability, 21–22, 34–36; and general defenses, 124–126, 142–144, 153, 155; and intentional offenses, 66–69; and negligence, 96–97, 99; and sentencing, 162–165, 167, 170–175; and strict liability, 112, 114
creativity, 7, 9, 20

de minimis, 123, 152–154
death penalty. *See* capital punishment
Deep Blue, 11
deontological morality, 18
Descartes, 1
deterrence, 76, 81, 157–159, 161, 166–168, 170, 172, 174
discipline, 150, 167
dreaming, 134
drone, 4, 104, 143, 151–152
duress, 120, 123, 140, 147–149, 155
dynamic modification, 10

evil, 25–26, 33, 44, 56, 64–65, 68; and necessity, 145–147; and negligence, 96
exemptions, 122–123
ex officio, 138
expert systems, 3–4, 8, 11, 13, 84, 93. *See also* general problem solver
external knowledge, 7–9, 24
eye for an eye, 157

factory robots, 13
factual element requirement (of criminal liability), 8, 29–32, 35; and general defenses, 118, 127, 136; and intentional offenses, 38–39, 41–46, 48, 57, 65, 72–73, 82; and negligence offenses, 85–90, 101; and strict liability offenses, 105–111
factual mistake, 72, 121, 123, 133, 135–137, 154
factual reality, 8, 10, 31–32, 45, 47; in general defenses, 120, 129, 131, 133–136, 154; and sentencing 160
fine, 26, 106, 158, 162–163, 165, 173–175
foreseeability, 58–64, 79, 100–101, 117
formal logic, 6, 8, 29

general intent, 32, 44, 57
generalization, 10–11, 84, 92, 100
general problem solver, 2, 6
good faith. *See bona fide*

heuristics, 3
Hobbes, 1
Hollywood, 15
homicide, 25, 32; and general defenses, 121, 131, 147, 154; intentional, 38, 42–44, 48, 75–76, 78–80, 82–83; negligent, 86, 94, 96, 98, 100–103; and sentencing, 166; strict liability, 117–119
humanoid, 14, 25
human-robot coexistence, 14–15, 17, 22

imminent danger, 141–142, 145, 148
imprisonment, 26, 142, 156, 162–165, 167–169, 171, 173, 175–176
inaction, 16–17, 28, 39–41, 87
incapacitation, 23, 157, 161–162, 166–172, 176
incarceration, 162, 164, 167, 169. *See also* imprisonment
indexing, 10
indifference, 47–49, 56, 58, 61–63
infancy, 28, 123–126
in personam, 122–123
in rem, 122, 140, 153
insanity, 28, 31, 65, 72, 121, 123, 126; as general defense, 128–130, 154–155
internal knowledge, 7–9
intoxication, 72, 123, 126, 130–133, 154–155
irresistible impulse, 129, 132

just deserts, 157–158, 160. *See also* retribution
justifications, 96, 112, 122–123, 140

Kant, 66

legal mistake, 123, 136–138
legal social control, 26–28
Leibniz, 1
lesser of two evils, 146–147, 149
loss of self-control, 39, 47, 123, 126–128, 154–155

machina sapiens, 1, 4–5, 7, 10–11, 13, 17–21, 55, 61, 68
machina sapiens criminalis, 1, 12–14, 17, 19, 21; and intentional offenses, 55, 61, 68, 82–83; and negligence, 101; and strict liability offenses 110–111, 117–118
machine learning, 4, 11; in general defenses, 125, 146, 149, 151; in intentional offenses, 62, 64, 75; in negligence offenses, 84, 92–95, 97, 100; and sentencing, 160–161, 171–173; in strict liability offenses, 114, 125, 146, 149, 151
mechanical computer, 1
mens rea, 32–33, *33*, 36–37; and general defenses, 120–121, 129–131, 135–136, 147; and intentional offenses, 44–49, 51–52, 56–57, 63–64, 67–69, 74, 79–80, 82, and negligence, 85–86, 89–91, 95–96, 98–102; and strict liability, 105–112, 114–117, *115*
mental element requirement (of criminal liability), 29, 31–41; and general defenses, 118–119, 121, 127–128, 135–137, 139; and intentional offenses, 44–45, 51, 62, 64–65, 67, 69, 72–77, 79–80, 82–83; and negligence, 84–87, 89, 95–101, 103–104; and sentencing, 156–157; and strict liability, 109–112, 114–115
monitoring, 4
moral accountability, 17–19, 65
morality, 18, 26, 65–66, 153

necessity, 120, 123, 140, 144–149, 155
negligence, 32–33, *33*, 36–38; and general defenses, 125, 135, 144; and intentional offenses, 45, 68, 72, 74, 79–80, 82; as offenses' type, 84–103; and strict liability, *115*, 115–118
nullity, 47

obedience, 151
objectivity, 90, 94
omission, 32–33, 36, 39–41, 86–87, 90

Pascal, 1
perception, 3, 49–52, 102, 131, 133, 136
police robots, 13
prediction, 10
prison guard robots, 13, 18
probable consequence liability: and general defenses, 138; in intentional offenses, 69, 75–81, 83; in negligence, 100–101, 103; in strict liability 116–117, 119
probation, 162–163, 165, 169–173
proportionality, 144
prostitution, 13
public service, 165, 167, 171–175
punishability, 19

rashness, 47–48, *49*, 56, 58, 61, 63–64
reasonableness, 24, 56, 144
recidivism, 167–168
recklessness, 28, 31, 48, *49*, 56, 58, 61–63, 82; and negligence, 85–86
rehabilitation, 122, 157, 159–162, 166, 168–172, 176

results (factual element component), 30–32, *31*; in intentional offenses, 38, 43–52, *49* 55, 57–60, 71; in negligence offenses 85, 87–88, *89*, 92, 97, 100; in strict liability offenses, 105, 108–109, *109*
retribution, 157–160, 166–170, 172, 174. *See also* just deserts

self-defense, 28, 65, 120–123, 140–142, 144–149, 154–155
semi-innocent agent, 69, 72, 74, 97–99, 114–115, *115*
sex with robots, 13
slavery, 69
society's absolute monopoly on power, 140, 144, 147
specific intent, 44, 47–48, 57, 59
strict liability, 32–33, *33*; and intention, 36–38, 45, 68, 74; and negligence, 85; offenses, criminal liability for, 104–119, *115*; and general defenses, 135, 137
substantive immunity, 123, 138, 140
superior orders, 123, 149–152, 155

teleological morality, 18, 65
thinking machine, 1, 4–5
Turing, 5–7, 14

ultra vires, 36, 71

vicarious liability, 35–36, 69–71
volition, 31–32, *33*, 45–49, 61–63, 88

Watson, 11
willful blindness, 52, 55–56

zoological legal model, 22–25